Research Schools on Number Theory in India

Purabi Mukherji

Research Schools on Number Theory in India

During the 20th Century

Springer

Purabi Mukherji
Gokhale Memorial Girls' College
Kolkata, West Bengal, India

ISBN 978-981-15-9622-3   ISBN 978-981-15-9620-9   (eBook)
https://doi.org/10.1007/978-981-15-9620-9

This Springer imprint is published by the registered company Springer Nature Singapore Pte Ltd.
The registered company address is: 152 Beach Road, #21-01/04 Gateway East, Singapore 189721,
Singapore

I dedicate this book to the memory of my father

Late Sushil Kumar Banerji

and my mother

Late Ranu Banerji

Foreword

It is with great pleasure that I write this foreword for the book, *Research Schools on Number Theory in India: During the 20th Century*, written by Dr. Purabi Mukherji. Since my expertise is neither number theory nor the history of science, I was quite apprehensive when Dr. Mukherji called to ask me if I would be willing to write a foreword for her book. After a little protest, however, I agreed.

I met Dr. Mukherji for the first time at the National Conference of Calcutta Mathematical Society in December 2015. At that first meeting, she told me that she was working on a project sponsored by the Indian National Science Academy (INSA), and we talked a little about the nature of this project. It was the start of a valuable friendship and we have talked to each other a lot since, trying to convince ourselves that the world around us perhaps will get a lot better one of these days and we may even live long enough to see it!

This is not the first book that Dr. Mukherji has authored. She co-authored a book, *History of the Calcutta School of Physical Sciences*, with Dr. Atri Mukhopadhyay. I was able to get hold of a copy of this book as soon as it was published and thoroughly enjoyed reading it. Given that I have no real sense of history of either science or anything else for that matter, I found this reading to be especially enlightening. Dr. Mukherji had called me some time ago telling me that for the INSA project, she was preparing a report on the research conducted on number theory in India during the twentieth century. I was very pleased to learn that this important project was in the hands of a very competent and knowledgeable person.

As I have said several times in many public forums, the research in number theory in India has been always of very high quality. I believe, it is for this reason, a very large number of bright young students have chosen to work in this area. Many of them today are counted among the leaders in this field. Unfortunately, there has been no serious account of the large body of work in the area of number theory during the twentieth century from around the country. For this reason, now that the INSA project has been turned into a book, I consider it to be a very timely and much needed addition to our literature on the subject.

Returning to the book itself—it was a fascinating and thoroughly engaging read. The organization of the book, from south to north followed by east to west, is very appealing. I was pleasantly surprised to see that there is a chapter on the applications of number theory to other areas. It has a complete list of all the research papers written during this period, which I am sure will be immensely useful to a student researching in this subject. Finally, as the author says and I agree, the book is written without using too much mathematical jargon to make it accessible to a general readership.

I intend to be coming back to this book ever so often. I am sure anyone who has some interest in the history of research in mathematics in India would enjoy reading this book as much as I did.

May 2020

Gadadhar Misra, FNA, FNASc, FASc
Professor and Chairman
Department of Mathematics
Indian Institute of Science
Bangalore, India

J. C. Bose, Fellow and Vice-President
Publications and Informatics
INSA, New Delhi, India

Preface

The great German mathematician Felix Klein had said "God is a geometer". Jacobi changed this and wrote "God is an arithmetician". Then came Kronecker and made the memorable comment: "God created the natural numbers and all the rest is the work of man". Number Theory, or the Theory of Numbers, has always held a special and unique position in the world of mathematics. Its fundamental place in the discipline has led to its sobriquet as "the queen of mathematics".

India had a long tradition of research in Number Theory since ancient times. Famous and eminent Indian mathematicians like Brahmagupta, Bhaskaracharya, Jayadeva, and others made notable contributions in this area. In the Western world, too, the all-time mathematical greats such as Euler, Gauss, Lagrange, Jacobi, and many others have made notable contributions in this area.

In India, research on Number Theory in modern times started with the advent of the iconic genius Srinivasa Ramanujan. His theories, conjectures, and questions on various topics of Number Theory inspired mathematicians both at home and abroad. From the early part of the twentieth century, inspired and influenced by Ramanujan, notable research was done in South India and Panjab. The establishment of Tata Institute of Fundamental Research (TIFR), in Bombay (now Mumbai), was a landmark event in the research activities related to Number Theory in the Indian context.

In this book, an attempt has been made to describe the gradual development of the major schools of research on Number Theory in South India, Panjab, TIFR, and the minor schools of Bengal and Bihar. A comprehensive bibliography of major number theorists of India during the twentieth century has been carefully compiled. Of course, the research contributions, which some of them have made after leaving India for good, have not been included.

A chapter has been devoted to the impact of the research made by Indian number theorists nationally and internationally. Finally, applications of some results developed by the indigenous number theorists in practical fields have been discussed. In the concluding chapter, a general discussion on the importance of Number Theory in the modern-day world of mathematics has been briefly discussed. A whole chapter has been devoted to the bibliographies of the number theorists whose works have been discussed in the text. It is valuable from the standpoint of historic documentation and will also serve the researchers in the field of Number Theory in their literature survey.

Since the book is written from the viewpoint of the "history of science", mathematical expressions and technical jargons have been avoided as much as possible. The language has been made lucid so that general readers with an attraction for scientific developments in India can easily understand the exposition.

Kolkata, India   Purabi Mukherji
January 2020

Acknowledgements

I wish to acknowledge and convey my grateful thanks to Indian National Science Academy (INSA), New Delhi, because this book is the outcome of the project titled "The Development of the School of Research on Number Theory in India During the twentieth Century", completed under the sponsorship of Indian National Commission for "History of Science", a wing of the INSA. I also convey my sincere thanks to the editorial board of INSA's journal, the *Indian Journal of History of Science* (IJHS), for permitting me to reproduce certain parts of my article on number theory published in the June 2019 issue of the journal.

I wish to sincerely thank Prof. M. S. Raghunathan, FRS, for his kind help in sending the photographs and also for giving his advice during the writing of the book. My grateful thanks are due to Prof. R. Balasubramanian, Prof. Sanoli Gun, and Prof. S. Ashok (IMSc, Chennai), Prof. A. Sankaranarayanan and Prof. C. S. Rajan (TIFR, Mumbai), Prof. S. D. Adhikari, Prof. R. Thangadurai, and Prof. D. Surya Ramana (HRI, Allahabad), Prof. Pradipta Bandyopadhyay (ISI, Kolkata), Prof. Parthasarathi Mukhopadhyay (RKM Residential College, Kolkata), and Prof. Pradip Majumdar (RBU, Kolkata) for their kind help in various matters related to the book.

Finally, I wish to thank the librarians and the library staff of Institute of Mathematical Sciences (IMSc), Chennai, Harish Chandra Research Institute (HRI), Allahabad, Tata Institute of Fundamental Research (TIFR), Mumbai, Indian Statistical Institute (ISI), Kolkata, Indian Association for the Cultivation of Science (IACS), Kolkata, and Calcutta Mathematical Society (CMS), Kolkata, for kindly entertaining me and helping me as and when I needed to consult books and journals. I offer my sincere thanks to my former student Miss Parama Paul for helping me with the proof reading and corrections.

Contents

Chapter 1
Historical Prelude and Introduction

Since the beginning of the recorded history, numbers have fascinated the human mind. In every known civilization, mathematics related to numbers was practiced. In ancient Egypt, the famous *Rhind* (or *Ahmes*) *Papyrus* was written in around 1700–1600 BC. They are safely stored in the British Museum in London. Information about the kind of mathematics practiced in ancient Egypt can be obtained from them. They deal with numbers as expressed by ancient Egyptian mathematicians. Cuneiform tablets recovered from Mesopotamia display evidences about the mathematics practiced there at the end of the third millennium BC. It confirms that arithmetic was quite developed at that time. In China, by the eleventh century BC, mathematics had been developed independently. They had developed number sets, comprising very large numbers and negative numbers, decimals and binary systems, and many more important theories and techniques. They were also quite proficient in calculating the value of pi (or π), division, and root extraction. Pascal's triangle, as known in the Western world, was known to the Chinese for at least three centuries even before Pascal was born (1303 CE). It is mentioned in the work of Chu Shih-Cheih.

Another major civilization was the ancient Indian civilization that prospered around the Indus Valley in India. Here, mathematics has a very rich history going back to at least or about 3000 BC. This tradition continued for centuries. The concept of zero as a number in its own right, as is now generally accepted internationally, was first developed in India. This is a seminal contribution. It should be specially noted that the transformation of zero from a simple placeholder to a number in its own right gives credence to the belief that the subcontinent had a prevalent highly developed mathematical culture. Apart from that the Indian contribution in the fields of arithmetic, negative numbers, geometry, trigonometry, and astronomy are also remarkable.

In India, during the Vedic period (approximately in 1200 BC), there is enough documentary evidence to prove that the mathematicians of those times were strong in dealing with arithmetical problems of different types. They were competent to handle arithmetic of fractions and surds and found good rational approximations

© The Author(s), under exclusive license to Springer Nature Singapore Pte Ltd. 2020
P. Mukherji, *Research Schools on Number Theory in India*,
https://doi.org/10.1007/978-981-15-9620-9_1

to irrational numbers like $\sqrt{2}$. In Vedic texts, numbers were usually expressed as combinations of powers of ten.

The *Shulba Sutras* written in 800 BC are the oldest geometrical treatises available in the subcontinent. The arithmetic content of these *Shulba Sutras* consists of rules for finding Pythagorean triples such as $(3, 4, 5)$, $(5, 12, 13)$, $(8, 15, 17)$, and $(12, 35, 37)$. Actually, Baudhayana gave a statement in terms of sides and diagonals of a rectangle which, later, the Greek mathematicians presented as the *Pythagorean theorem*. The triples mentioned above arose out of a religious ritual, where every Hindu home was required to have three fires burning at three different alters of square, circular, and semicircular shapes, having the same area. These conditions led to certain types of "Diophantine" problems.

Around the seconnd century BC, mathematicians of India were aware of the "binary system", as one may see in *Pingala Chanda Sutra*. The concept of modern-day Pascal's triangle is also found in *Pingala Chanda Sutra* 1500 years before the Chinese invention. Later, over the centuries, many Hindu and Jaina mathematicians made notable contributions in the fields of arithmetic.

Aryabhata 1 (476–550 AD) computed the value of π correct to four decimal places. He also clearly stated that it was but an approximate value only. He also dealt with continued fractions and discovered algorithms for finding approximate square roots and cube roots. The so-called "generalized Euclidean algorithm" related to g.c.d. is actually due to him.

The famous Hindu mathematician Brahmagupta (sevenh century AD) in *Brahmasphutasiddhanta* initiated rules for working with negative numbers. The rules for the four basic arithmetical operations namely, addition, subtraction, multiplication, and division, were initiated by him. He was the first mathematician to state formally that any number added to its own negative gives rise to zero. The solution method of the so-called "Pell's equation" is now internationally attributed to Brahmagupta and his famous principle of *Bhavana,* which was later rediscovered in Europe by Euler.

Mahavira Acharya (800–870 AD) and Virasena (eighh century AD) were Jaina mathematicians who dealt with 2-adic orders. Sridhara (870–930 AD) who lived in West Bengal authored a treatise named *Pati-ganita*. There he dealt with fractions and gave eight rules for operations involving zero. He also initiated methods of summation of different arithmetic and geometric series.

Aryabhata II (920–1000 AD), in his astronomical treatise named *Maha-Siddhanta,* discussed in detail the methods of numerical mathematics (*ankganit*) and solution of indeterminate equations (*kuttaka*).

Bhaskara II (1114–1185 AD) made substantial contributions to arithmetical and geometrical progressions. He called a fraction whose denominator is zero *khahara* and gave an allegorical description of this entity as comparable with the infinite immutable God, from whom everything is created without reducing Him, while everything merged within Him after destruction but without increasing Him. In modern mathematical terminology, this corresponds to "division by zero is infinity". He was also familiar with a positive number which had two square roots. He worked on surds and developed methods for solving various kinds of equations. His technique of solving the general form of "Pell's equation" and the general indeterminate quadratic equation using the *chakravala* method (originally developed by Jayadeva of about 200 years earlier) is highly acclaimed. He also did commendable work on indeterminate cubic, indeterminate quartic, and indeterminate higher-order polynomial equations. He was successful in computing the value of π correct to five decimal places.

So, it is evident that starting from very ancient times till the twelfth century, Indian mathematicians had an unbroken tradition of working with numbers and related problems. This tradition was successfully followed and much furthered by the Kerala School of Madhava of Sangamagrama from the fourteenth century to the seventeenth century.

Coming to more recent times of all branches of mathematics, the greatest mathematical genius of the nineteenth century, C. F. Gauss (1777–1855) hailed "number theory" as the queen of mathematics. The natural and pertinent question would be why this particular branch of mathematics is so highly adulated. Probably, the reason for this glorification of the said topic is because of the fact that in this particular branch of mathematics, it is possible to ask questions in very simple terms: easily understandable by anyone but not so easily answerable by all. In contrast, in any other branch of mathematics, asking a really meaningful, non-trivial question, a long list of definitions is a mandatory prerequisite.

The "theory of numbers" (or number theory) is a branch of mathematics that deals with properties and relationships of numbers, especially the positive integers. Number theory can be broadly classified as "analytic" and "algebraic". "Analytic number theory" is a branch of number theory that uses methods and theories from mathematical analysis to solve problems about the integers. On the other hand, "algebraic number theory" studies algebraic structures related to algebraic integers.

From the world perspective, the twentieth century has been witness to an explosion to number theory-related research work. Along with classical and analytic number theory, scholars started exploring specialized subfields, such as algebraic number theory, geometric number theory, and combinatorial number theory.

Some of the greatest mathematical minds of modern times, such as Leonhard Euler (1707–1783), Carl F. Gauss (1777–1855), J. L. Lagrange (1736–1813), P. De Fermat (1601–1665), C. G. J. Jacobi (1804–1851), E. Waring (1734–1778), S. Ramanujan (1887–1920), P. Erdös (1913–1996), and many others have worked on various problems related to number theory.

Since this book is devoted to the development of the "schools of research" on number theory in India during the twentieth century, the focus will be the discussion on Indian number theorists of the time and their contributions in developing such centers of research in the field.

Chapter 2
Indian Schools of Research on Number Theory

The development that took place in the research activities on number theory in India during the twentieth century is quite vast and a chronological study of the same has been undertaken during the course of the present exercise. The different schools of research developed during the period have been identified, and detailed discussions on the notable contributions of famous number theorists of India have been done. The research schools that developed during the first half of the twentieth century have been dealt with first. The ones that developed roughly during the second half of that century have been discussed later.

The two major schools that developed in the first half of the twentieth century are, respectively, South Indian School and the Panjab School of research on number theory. A small group of researchers from Bengal and Bihar also did carry out some research in the said discipline during that time. It would of course be pertinent to admit at the outset that compared to the South Indian and Panjab Schools of research on number theory, the contribution of the Bengal–Bihar schools was pretty meager. But for the sake of historical records, it is of some relevance.

In the second half of the twentieth century, the Panjab School continued to flourish. But the most outstanding center of research on number theory was established in the newly founded Tata Institute of Fundamental Research (TIFR) in Mumbai. Detailed discussions on all these centers have been done and they form the main text of the book. The analysis is made on a chronological basis to unravel the historical significance of the whole exercise.

© The Author(s), under exclusive license to Springer Nature Singapore Pte Ltd. 2020
P. Mukherji, *Research Schools on Number Theory in India*,
https://doi.org/10.1007/978-981-15-9620-9_2

2.1   South Indian School of Research on Number Theory (1910–1950)

Here, the discourse is about the development of number theory-related research work that was initiated for the first time by Indian mathematicians born in the southern part of India. The first half of the twentieth century was a witness to phenomenal growth in the field, and brief discussions about the main contributors in the field are presented below.

2.1.1   Srinivasa Ramanujan (1887–1920)

Any discussion related to number theory has to start with Srinivasa Ramanujan. He was one of the world's finest mathematicians and his contributions to number theory unambiguously make him the greatest number theorist of the world. He has been the fountainhead of inspiration to generations of mathematicians in India as well as the world. He has been a key figure who has drawn number theorists both nationally and internationally to prove his conjectures and theorems and solve problems posed by him directly or indirectly connected to number theory. In a general way, too, S. Ramanujan was a major influence in the field of scientific research in India during the early years of the twentieth century. A remark by the Nobel Laureate S. Chandrasekhar is quite relevant in this context. He wrote:

> Let me turn to the role of Ramanujan in the development of science in India during the early years of this century (twentieth century). Perhaps the best way I can give you a feeling for what Ramanujan meant to the young men going to schools and colleges during the period 1915–1930 is to recall the way in which I first learnt of Ramanujan's name....

Then he related how he came to know about Ramanujan's premature death and his great achievements during his stay at Cambridge. What impressed Chandrasekhar most was that even though Ramanujan was brought up in abject poverty and in a scientifically sterile environment, he had succeeded in conquering all the adversities and established himself as one of the most original mathematicians of the century. Chandrasekhar felt that those facts were more than enough to encourage aspiring young Indian students to break the bonds of their intellectual confinement and establish themselves at a spectacular height as Ramanujan had done. Chandrasekhar further commented:

> The twenties and thirties were a period when young Indians were inspired for achievement and accomplishment by the men they saw among them.

The life story of Srinivasa Ramanujan is fairly well known to the scientific community in India. Still, for the completeness of historical facts, a brief outline of the life history of this iconic Indian mathematician is given below.

Ramanujan (Fig. 2.1) was born on December 22, 1887, at Erode in Tamil Nadu

Fig. 2.1 Srinivasa
Ramanujan (1887–1920)

in a Brahmin family. His parents belonged to the financially weaker section of the
society and had to work hard to make both ends meet. At the age of 7, he was
sent to a primary school situated in the town of Kumbakonam. He stood first in the
Tanjore district primary examination held in 1897. This helped him to get a half-free
studentship in the Town High School at Kumbakonam. He studied there from 1898 to
1903. In 1904, he passed the Matriculation examination, conducted by the University
of Madras.

Right from his school days, Ramanujan showed extraordinary intuition and
astounding proficiency in various branches of mathematics such as arithmetic,
algebra, geometry, number theory, and trigonometry. The senior mathematics teacher
of the school was so confident about Ramanujan's mathematical prowess that every
year he entrusted young Ramanujan with the task of preparing a conflict-free
timetable for the whole school. While still in his fourth form, Ramanujan mastered
Loney's *Trigonometry* (Part II). During the school years, Ramanujan won many
prizes for his outstanding performances in mathematics.

To augment the meager family income, every year Ramanujan's mother took in
a few students as boarders. These undergraduate students noticed the young boy's
unusual fascination for mathematics and out of kindness for him, they gave him an
elementary introduction to all branches of mathematics. In 1903, with the help of
friends from the local government college, Ramanujan was able to procure G. S.
Carr's *A Synopsis of Elementary Results*, a book on Pure mathematics. The book,
published in 1896, contained propositions, formulae, and methods of mathematical
analysis with brief hints. Ramanujan started working tirelessly to solve the problems

given in the book. The more he worked, the more he became involved with mathematics. A biographer of Ramanujan, Prof. P. V. Seshu Aiyar, in his edited book has
stated:

> It was this book which awakened his genius. He set himself to establish the formulae given
> therein. As he was without the aid of other books, each solution was a piece of research so
> far as he was concerned.[1]

The fact is that while proving formulae from Carr's book, Ramanujan discovered
many new ones. The self-taught genius thus laid for himself the foundations for
pursuing higher mathematics. During 1904–1907, he also started noting down the
new formulae in his now famous *Notebook*. After completing school education,
Ramanujan won a junior scholarship for his proficiency in mathematics and English
composition. This enabled him to join the First Examination in Arts (FA) course at
the Government Arts College at Kumbakonam. Due to his obsessive preoccupation
with mathematics, he neglected the other subjects of the course and failed. As a
consequence, he lost his scholarship and was not promoted to the senior FA class
either. The next year, in 1906, Ramanujan for the first time left his hometown and
moved over to Madras and got admitted into a college there to continue the FA.
Due to ill health, he had to discontinue his studies. In 1907, he appeared for the FA
examination as a private candidate. Though he secured full marks in mathematics,
he failed in the other subjects. That was the end of his formal education.

In spite of endless struggle with abject poverty, Ramanujan still maintained an
unfailing commitment to mathematics. Those were years of great adversity for him.
Despite the stresses and strains of day-to-day existence, Ramanujan continued noting
down the mathematical results conjectured or discovered by him in *Notebooks*. Years
later after Ramanujan was no more, his mentor and friend Prof. G. H. Hardy (1877–
1947) had lamented and wrote:

> The years between 18 and 25 are the critical years in a mathematician's career.... During his
> five unfortunate years (1907–1912), his genius was misdirected, sidetracked and to a certain
> extent distorted.[2]

Finally, from 1910 onwards a few lovers of mathematics in Madras (present-day
Chennai) took notice of this exceptionally talented mathematician. The earliest
contributions of Ramanujan were communicated by Prof. Seshu Aiyar to the Journal
of the Indian Mathematical Society in the form of "questions". They were published
in 1911. Ramanujan's first research paper on number theory also appeared in the
same issue of the journal and was titled *"Some Properties of Bernoulli Numbers"*
[Sect. I, (SR. 1)]. In the paper, he stated eight theorems concerning the arithmetical
properties of Bernoulli numbers of which he proved three, two were stated as corollaries, and three were just conjectures. In 1912, on the advice of Prof. Seshu Aiyer,

---

[1]G. H. Hardy, P. V. Seshu Aiyer and B. M. Wilson (Eds.).*Collected Papers by Srinivasa Ramanujan.*
New York: Chelsea. 1962.

[2]G. H. Hardy. *Ramanujan: Twelve Lectures on Subjects Suggested by His Life and Works.* New
York: Chelsea. 1940.

Ramanujan communicated some of his mathematical results to Mr. G. H. Hardy who was the Cayley Lecturer in mathematics at the University of Cambridge. Hardy was a world-renowned mathematician and a Fellow of the Royal Society of London. This first letter of Ramanujan to Hardy dated January 16, 1913, is historically very signifi-cant. After initial hesitation, Hardy finally decided to invite Ramanujan to Cambridge. Hardy's reply dated February 8, 1913, was the beginning of Ramanujan's recognition by the mathematicians of the Western world.

With financial assistance from the University of Madras and help from friends and well-wishers, Ramanujan finally reached London on the April 14, 1914. On April 18, he reached Cambridge and was soon admitted by Prof. Hardy to Trinity College there. Though Ramanujan did not have much formal education and had access to only Carr's synopsis and perhaps to a book on Jacobi's elliptic functions, according to the historian J. R. Newman:

> He arrived in England abreast and often ahead of contemporary mathematical knowledge. Thus in a lone mighty sweep, he had succeeded in recreating his field, through his own unaided powers, a rich half century of European mathematics. One may doubt whether so prodigious a feat had ever before been accomplished in the history of thought.[3]

Then started Ramanujan's brief but brilliant research career, and he made valu-able contributions to various branches of mathematics including number theory. Ramanujan, as has already been pointed out, was practically a self-taught natural genius. G. H. Hardy, the eminent British mathematician who discovered and was responsible for taking him to Cambridge wrote:

> … He worked far more than the majority of the modern mathematicians, by induction from numerical examples; all his congruence properties of partitions, for example, were discovered in this way. But, with his memory, his patience, and his power of calculation, he combined a power for generalization, a feeling for form, and a capacity for rapid modification of his hypotheses that were often startling, and made him in his own field, without a rival in his day.[4]

Professor Hardy further added:

> I do not think now that this extremely strong language is extravagant. It is possible that the great days of formula are finished, and that Ramanujan ought to have been born 100 years ago; but he was the greatest formalist of his time.

The above statements clearly indicate why Ramanujan was and still remains an iconic source of inspiration to research workers not only in India but also all over the world. Ramanujan's work on number theory is of very large expanse. He has made notable contributions in various topics such as modular equations, theory of partitions, highly composite numbers, Diophantine equations, probabilistic number theory, and mock theta functions. In the Indian context, it would be fair to say

---

[3] J. R. Newman: Article titled "Srinivasa Ramanujan" in the book *Mathematics in the Modern World*, W. H. Freeman & Company, 1968.
[4] G. H. Hardy. *Ramanujan*. Cambridge University Press, (1940), 14.

that most of the mathematicians who achieved distinction during the three or four decades following Ramanujan were directly or indirectly inspired and influenced by his examples.

Attempts would be made to briefly discuss his most important works on theta functions, elliptic functions, modular equations and modular forms and their connection with number theory. Special attention would be focused on certain arithmetic functions on which Ramanujan made very valuable contributions. They are partition functions $p(n)$ and Ramanujan's tau function, $\tau(n)$. Most of these works are found in Ramanujan's published papers on $p(n)$ and $\tau(n)$ [Sect. 2.1, (SR. 10, 20, 22, 23)]. Earlier, he had noted them down in the later chapters of his second *Notebook*. It is quite well known that Ramanujan left very few proofs for the conjectures that he wrote down in his *Notebooks* including the *Lost Notebook*.

One may safely say that most of Ramanujan's important research works on number theory came out of q-series [1] and theta functions [2]. Theta functions are the fundamental bedrocks in the theory of elliptic functions [3]. The ordinary definition of elliptic functions in very simple terminology may be stated as meromorphic [4] functions with two linearly independent periods over the real numbers. But Ramanujan never used the concept of double periodicity. In "Notebooks 1 and 2", Ramanujan has devoted 70% of the space to develop formulae of elliptic function. Nowhere in "Notebooks", the double periodicity of these functions has been mentioned. In fact, even if a function occurring in his notes displayed double periodicity, he did not take it into consideration.

In "Notebooks", Ramanujan has used his own peculiar unique notation and has developed in detail the theta function of Jacobi. But the famous parameter, q, is totally absent in *Notebook*. Even though, in his "Notebooks", Ramanujan has given plenty of examples of modular equations of several degrees, the term "modular equation" is absent there.

In modern-day higher mathematics, it has been seen that elliptic functions can be expressed as inverse functions of elliptic integrals. Jacobi, in 1829, originally developed the theory of theta functions. Theta functions are an important class of functions closely related to elliptic functions. Experts on Ramanujan feel that since before going to Cambridge, he had not seen any standard book on elliptic functions, Ramanujan had used his own notations and was unaware of the fundamental property of double periodicity of the said functions.

It may be noted here that in Chap. 16 of Ramanujan's *Second Notebook*, which was published by the Tata Institute of Fundamental Research, Bombay, in 1957, the iconic mathematician has given a complete list of identities on theta functions and related q-series. While developing the theory of theta functions, he also commenced his study of elliptic functions. In the said chapter, the first 17 sections primarily contain his studies on q-series, while the later 22 sections are devoted to a thorough development of the theory of theta functions. Altogether, in this particular *Notebook*, there are 135 theorems, corollaries, and examples. According to noted number theorist Prof. R.P. Agarwal:

...which constitutes this chapter, and form one of the most elegant illustrations was the essential beauty and perfection of classical analysis par excellence.

In 1985, the American Mathematical Society published *Memoir* on Ramanujan.[5] In this publication, C. Adiga, B. C. Berndt, S. Bhargava, and G. N. Watson gave a brief survey of Chap. 16, which has been discussed in the previous paragraph. They have successfully proved all the 135 identities on theta functions and q-series as given by Ramanujan. It would be pertinent to state here that much more significant and deep results on theta functions have been stated without proofs by Ramanujan in his *Lost Notebook*. These identities could be further generalized in the context of basic hypergeometric series [5] and branched out to allied areas. The studies carried out by these mathematicians opened up new avenues of research in these areas.

Ramanujan's *Lost* Notebook

During the last few months of his life in 1920, Ramanujan was bed-ridden in Madras. In that condition, he used to write down many formulae. After his death, his wife Janaki Ammal collected these sheets and through Mr. Dewsbury, the Registrar of Madras University, sent them to Prof. Hardy in Cambridge. Professor Hardy requested Prof. G. N. Watson (1886–1965) to sort out this last manuscript of Ramanujan. Hardy felt that the work was more of the type that Watson did. Unfortunately, before Watson could finish the job assigned to him, he too expired.

In 1976, American mathematician G. E. Andrews who was then a young University of Wisconsin visiting professor bound for a one-week conference in France made a side trip to Cambridge. On a suggestion from a colleague, he had decided to rummage through some papers left behind by Prof. G. N. Watson.

Professor Watson was a Fellow of the Royal Society. After his death, the Royal Society asked J. M. Whittaker, son of one of Watson's collaborators, to write his obituary. Whittaker sought the permission of Mrs. Watson for looking into her late husband's papers. Mrs. Watson invited him to lunch and took him to late Prof. Watson's study, where Whittaker found the floor of the fair-sized study covered by papers to a depth of about a foot, all jumbled together, and were planned to be burnt up in a few days. Whittaker made several dips in the heap and by an extraordinary stroke of luck brought up the Ramanujan material.

This "material", totalling about 140 pages, was part of a batch of papers which Dewsbury had sent to Hardy in 1923. Whittaker collected the papers and passed them on to Robert Rankin. He in turn in 1968 handed them to Trinity College, Cambridge.

Andrews while visiting the library of the Trinity College accidentally found the box containing more than 100 papers in Ramanujan's own handwriting lying unedited. Within a few minutes, he realized that there were inscriptions related to mock theta functions. Naturally, they were generated during the last year of his life in India. He requested the library authorities to photocopy them and send them to him to the USA. Later, after returning to the USA, on close and prolonged inspection, he found that there were over 600 formulae listed one after another, without any proofs.

---

[5]American Mathematical Society, Vol. 53, No. 315.

Andrews meticulously studied them and gave proofs of the most amazing identities contained there. After that, he very correctly termed these papers as *Ramanujan's Lost Notebook*. In a section of this "lost notebook" are formulae for theta functions, false-theta functions, and mock theta functions.

Ramanujan and Elliptic Functions

It is to be noted that the development of elliptic functions by Ramanujan is purely algebraic. He did not use the Cauchy theory. The Fourier double integral theorem had been used by him only occasionally. It would be relevant to mention here that Jacobi and his student Borchardt developed and established the details in a purely algebraic way. But their method was rather complicated, whereas the method used by Ramanujan was very simple.

In his *Notebook* 2, Chap. 14, Ramanujan in a random fashion had stated a number of formulae of the theory of elliptic functions. At the very outset, one comes across the imaginary transformation of the Dedekind η-function and the Eisenstein functions. It is rather strange that no elliptic function is seen in this chapter. Perhaps, the only possible explanation for this kind of presentation could be that since Ramanujan was an expert in Fourier-integral and Mellin-integral transformations, he employed this mastery to obtain imaginary transformation formulae of the elliptic transcendental function. To support this view, it may be worthwhile to look into Ramanujan's paper entitled "Some definite integrals".[6] In that paper, he has shown how effectively he can use these classical transformations. However, in Chap. 14 of *Notebook* 2, he has used very elementary methods. He has developed circular functions such as cot z and cosec z, where the involved series are absolutely convergent. In Example 18 of this chapter, Ramanujan has clearly and fully explained the method he has used. In Example 19, in the same chapter, he has given the expansion of $\pi^2 xy \cot(\pi x) \coth(\pi y)$ in a simple way derived by him. Ramanujan himself has remarked that such expansions can also be obtained by the method of "residue calculus". In this context, comments made by two eminent number theorists seem to be worth quoting.

As regards Ramanujan's work on elliptic and modular functions [6], G. H. Hardy commented[7]:

> ...that both the profundity and limitations of Ramanujan's knowledge stand out sharply. Ramanujan never professed to have made any major advance in the theory of elliptic functions, and it seems that he must have learnt the fundamentals of the theory, so far as he was interested in them, from books. There is a sharp contrast between this attitude and his attitude about the theory of primes, where he certainly regarded all the results as his own. He never writes as if he had invented theta functions or modular equations, though he sets out a whole theory of them in a language of his own.

Hardy again commented:

> And there is some of his work, mostly in the theory of elliptic functions, about which some mystery still remains; it is not possible after all the work of Watson and Mordell, to draw

[6] *Journal Indian Mathematical Society*, 11, (1919), 81–87.
[7] *Ramanujan.* Cambridge University Press. (1940).

the line between what he may have picked up somehow and what he must have found for himself.

E. Littlewood remarked:

Ramanujan somehow acquired an effective complete knowledge of the formal side of the elliptic functions.

A survey of Ramanujan's second "Notebook" reveals that in Chaps. 16–21, he has considered q-series and θ-functions; the fundamental properties of elliptic functions; Jacobi's elliptic functions; modular equations having degrees 3, 5, 7, and associated θ-function identities; modular equations of higher and composite degrees; and Eisenstein series.

Ramanujan's Modular Equations
Special discussions are necessary in this area in order to correctly evaluate Ramanujan's contributions. If one goes back to the earliest appearance of these equations, then the name of the first mathematician that comes to mind is A. M. Legendre (1752–1833). In 1825, in his treatise in French titled "Traile des fonctions elliptiques et des integrals euleriennes",[8] he first developed his modular equation of degree 3 given by

$$(\alpha\beta)^{1/4} + [(1 - \alpha)(1 - \beta)]^{1/4} = 1.$$

In the century that followed, many mathematicians like A. Cayley (1821–1895), A. Enneper (1830–1885), R. Fricke (1861–1930), C. G. J. Jacobi (1804–1851), F. Klein (1849–1925), G. N. Watson (1886–1965), and many others made important additions to the long list of modular equations. But B. C. Berndt, a famous mathematician and an expert on Ramanujan, claimed with authority and commented:

However, the mathematician who discovered far more modular equations than any of these mathematicians was Ramanujan who constructed over 200 modular equations.

Before proceeding to England, while still in India, Ramanujan had started working on modular equations. However, his first paper in this area titled "Modular equations and approximation to π"[9] was the first of his research papers to be published during his stay in England. In later years, too, he repeatedly worked on the same kind of mathematics. Unfortunately, most of Ramanujan's works on modular equations are in fragments, and many of them do not have any proofs. They have been published with his last "Notebook". How Ramanujan could have obtained these equations would always remain a mystery. Using the theories of modular form, present-day modern mathematicians have been successful in proving some of them.

---

[8]Paris, Huzard-Courciers (1825–1828) in three volumes. [Microform, Readex Microprint Corporation, New York, 1970].
[9]Section 1 (SR. 4).

In Chap. 20 of his second "Notebook", which was published by TIFR in 1957, it is seen that there Ramanujan has taken up another type of modular equation, which is known as "mixed" modular equation or modular equation of composite degree. A mixed modular equation may be defined as follows.

Mixed Modular Equation

If $K, K', L_1, L'_1, L_2, L'_2, L_3, L'_3$ denote complete elliptic integrals of the first kind associated in pairs with the moduli $\alpha^{1/2}, \beta^{1/2}, \gamma^{1/2}$ and $\delta^{1/2}$, and their complementary moduli and again if n_1, n_2, and n_3 are positive integers such that $n_3 = n_1 n_2$, and it is supposed that the equations $n_1 K'/K = L'_1/L_1$, $n_2 K'/K = L'_2/L_2$, and $n_3 K'/K = L'_3/L_3$ hold, then a mixed modular equation is a relation between the moduli $\alpha^{1/2}$, $\beta^{1/2}, \gamma^{1/2}$, and $\delta^{1/2}$ that is induced by the above relations. In such a situation, β, γ, and δ are of degrees n_1, n_2, and n_3, respectively.

It is now known that Ramanujan stated "mixed modular equations" for 20 distinct triples of positive integers n_1, n_2, and n_3. According to Prof. Bruce Berndt, no one other than Ramanujan has worked in the area of "mixed modular equations". It may be noted that the vastness of Ramanujan's findings in the theory of modular equations is awesome. For some values of n, Ramanujan has offered more than 10 different modular equations. As it is, the creation of modular equation is very complex. So, it is remarkable that in the vast expanse of Ramanujan's contributions to modular equations, there are only a couple of erroneous ones. In contrast to the modern sophisticated approaches to modular equations, Ramanujan's approaches were more "natural". This certainly calls for more extensive study and research on the subject to understand and appreciate Ramanujan's work better.

Apart from its theoretical aspects, modular equations have very important applicability as well. D. H. Lehmar, in his paper titled "Properties of the coefficients of the modular invariant $J(\tau)$",[10] derived congruential properties of the Fourier coefficients of $J(\tau)$ by using the modular equations of Klein's absolute invariant $J(\tau)$. Again C. Hermite in his two research papers written in French and published in 1859[11] and in 1908[12] demonstrated how using modular equations of degrees 5 and 7 general quintic and septic polynomials could be solved.

In a tribute to Ramanujan, noted number theorist Prof. K. Alladi wrote[13]:

Ramanujan's mathematics remains youthful even in the modern world of the computer. His modular equations were used by Canadians Jonathan and Peter Borwein to calculate π (the ratio of the circumference of a circle to its diameter) to several million decimal places. The Borweins showed that these modular equations produce efficient algorithms to obtain approximations to π and other numbers. More recently, Ramanujan's transformations for elliptic functions were used by David and Gregory Chudnovsky to produce very rapidly convergent algorithms to compute π; in fact the Chudnovskys have now calculated π to the order of about a billion digits!.

[10] *American Journal of Mathematics*, 64, (1942), 488–502.

[11] *C. R. Acad. Sci.* (Paris), 48, (1859), 940–947, 1079–1084, 1095–1102: 49, (1859), 16–24, 110–118, 141–144.

[12] Tome II, Gauthier–Villars, (Paris), (1908), 5–12, 38–82.

[13] *Ramanujan's Place in the World of Mathematics.* Springer, India, (2013).

Compared to the prevalent older methods, Ramanujan's modular equations are very elegant. In this matter, G. N. Watson's comment is noteworthy. He wrote:

> When dealing with Ramanujan's modular equations generally, it has always seemed to me that knowledge of other people's work is a positive disadvantage in that it tends to put one off the shortest track.

Another comment by Bruce Berndt is also quite relevant in this context. Berndt wrote:

> Ramanujan, to our knowledge, has not explicitly stated Schröter's formulae in any of his published papers, notebooks or unpublished manuscripts. But, it seems clear, from the theory of theta functions that he did develop that Ramanujan must have been aware of these formulae or at least of the principles that yield the many special cases that Ramanujan doubtless used. However, the majority of Ramanujan's modular equations do not appear to be results of Schröter's formulae. We conjecture that Ramanujan knew other general formulae involving theta-functions, which are still unknown to us, and which he used to derive further modular equations. In particular, we think that Ramanujan derived an unknown general formula involving quotients of theta-functions.

G. N. Watson in his paper titled "Ramanujan's notebooks",[14] had forcefully asserted and wrote:

> A prolonged study of his modular equations has convinced me that he was in possession of a general formula by means of which modular equations can be constructed in terrifying numbers.

Watson also felt that Ramanujan's "general formula" is Schröter's most general formula. There is a general and common feeling among most experts on Ramanujan in this regard that most probably he knew other formulae or analytic techniques which have escaped their attention.

In this context, it may be mentioned that Ramanujan's "modular identities" are actually identities between modular forms of a given type. He wrote them down systematically in his "Notebooks", including the "Lost Notebook" (which has already been mentioned and discussed in detail). These types of identities have been applied in diverse areas, congruence for partition functions being one of them. Another point to be specially noted is that although Jacobi, Cayley, and others mathematicians made extensive studies about elliptic integrals, Ramanujan was the first to discuss concrete elliptic integrals associated with modular curves of a small conductor. Another remarkable fact is that Ramanujan was aware of the dichotomy between elliptic integrals of elementary types and other difficult types, much before Hecke.

Ramanujan's τ-Functions

The next paper which Ramanujan published after reaching England was titled "On certain arithmetical functions".[15] In this modestly titled paper, Ramanujan introduced the concept of the τ-function. This is an integer-valued function on natural numbers.

---

[14] *Journal London Math. Soc.*, 6, (1931), 137–153.

[15] Section 1, (SR. 10).

It first came to be known as part of an "error term" while counting the number of ways that a number could be written as a sum of 24 squares. Ramanujan, however, was convinced that the function should be investigated as it had a lot of importance. During the twentieth century, a very significant theme of mathematics resulted from Ramanujan's conviction.

The τ-function is defined by

$$g(x) = x\left[(1-x)(1-x^2)(1-x^3)\cdots\right]^{24} = \sigma_r(n)$$

$$= x \prod_{n=1}^{\infty} (1-x^n)^{24}$$

$$= \sum_{n=1}^{\infty} \tau(n)x^n.$$

Ramanujan calculated and tabled the values of $\tau(n)$ up to $n = 30$ and showed that $\tau(p) \equiv 0 \pmod{p}$ for $p = 2, 3, 5, 7, 23$. For these primes, $\tau(pn) \equiv 0 \pmod{p}$ for every n. He was the first mathematician to notice interesting congruence properties of the τ-function such as $\tau(n) \equiv \sigma_r(n) \pmod{691}$, where $\sigma_r(n)$, the sum being all over d dividing n. The prime number 691 is called the *Ramanujan Prime Number*.

Ramanujan in his research paper mentioned above had formulated some fundamental conjectures. He stated that

(i) $\tau(n)$ is multiplicative, which mathematically means
(ii) $\tau(m)\,\tau(n) = \tau(mn)$, where m and n are positive integers and relatively prime
(iii) If p is any prime and n is any integer exceeding 1, then $(n \geq 1)$
(iv) $\tau(p^{n+1}) = \tau(p)\,\tau(p^n) - p\tau(p^{n-1})$
(v)   For each prime p, $|\tau(p)| \leq 2p^{11/2}$.

However, Ramanujan did not give any proofs to his conjectures.

The first two conjectures were proved by L. J. Mordell in his famous paper titled "On Ramanujan's empirical expansion of modular functions".[16] E. Hecke further generalized this proof in his two papers written in German.[17]

Ramanujan's conjectures and these publications by Mordell and Hecke led to the beginning of one of the most important eras in the theory of modular forms. Ramanujan had computed the first few values of τ-functions, such as $\tau(1) = 1$, $\tau(2) = -24$, and $\tau(3) = 252$. He had shown that $|\tau(n)| \ll n^7$ and had generalized the third conjecture further as $|\tau(n)| \leq d(n)\,n^{1/2}$, where $d(n)$ denotes the number of positive divisors of n. This result is known as *Ramanujan's conjecture*. Later, this was named as *Ramanujan's hypothesis* by G. H. Hardy. In this hypothesis, Ramanujan had actually conjectured a specific upper bound for the tau function. It was one of the most famous among his unproved conjectures. In 1939, Robert Rankin had in a paper, published in the *Proceedings of the Cambridge Philosophical Society,* obtained an

---

[16] *Proceed. Camb. Philos. Soc.*, 19, (1917), 117–124.
[17] *Math. Annals.*, 114, (1937), 1–28 and 316–351.

upper bound for the tau function and it was pretty close to Ramanujan's hypothesis. But finally in 1970, Pierre Deligne used sophisticated methods of algebraic geometry to fully solve *Ramanujan's hypothesis.* His detailed derivation was published in his paper titled "La conjecture de Weil I",[18] which was written in French. For this pathbreaking work, Deligne was awarded the Fields Medal in 1978. It is probably well-known that the Fields Medal is the equivalent of the Nobel Prize in mathematics.

In this context, a statement by another Fields medallist Prof. Atle Selberg is worth recalling. In 1988, during a talk at the Tata Institute of Fundamental Research, Bombay, he said:

> ... that a felicitous but unproved conjecture may be of much more consequence than the proof of many a respectable theorem. Ramanujan's recognition of the multiplicative properties of the coefficients of modular forms that we now refer to as cusp forms and his conjectures formulated in this connection, and their later generalization, have come to play a more central role in the mathematics of today, serving as a kind of focus for the attention of quite a large group of best mathematicians of our time.[19]

The proceedings of the centenary conference, held at the University of Illinois at Urbana-Champaign, in June 1987 were published as *Ramanujan Revisited* by the Academic Press, San Diego, in 1988. The volume was edited by a host of world-class mathematicians who were experts on Ramanujan's work as well. They were G. E. Andrews, R. A. Askey, B. C. Berndt, K. G. Ramanathan, and R. A. Rankin. In that volume, in a paper titled "The Ramanujan τ-function", R. A. Rankin wrote:

> The τ-function provides a good illustration of Ramanujan's insight and ingenious ability to find new interesting facts in areas neglected by other mathematicians. In several cases, many years have elapsed before subjects that he founded have been studied by others and have revealed their richness and importance.

In the same conference, M. Ram Murty in his paper with the same title as that of Rankin's remarked:

> Ramanujan was the first to foresee the arithmetical significance of $\tau(n)$, for he was the first to investigate the divisibility properties of these coefficients. From the works of Shimura, Serre and Deligne, it is now known that the decomposition laws of primes in certain non-solvable extensions of Q are given by divisibility criteria for the τ-function. The first indications of a general reciprocity law and non-Abelian class field theory lie hidden in the divisibility properties of Fourier coefficients of cusp forms.

Ramanujan's Contributions in the Theory of Partitions and the Additive Theory of Numbers

The idea of numbers and magic squares had fascinated Ramanujan from the early student days. His involvement with the theory of partitions is probably a natural consequence of that enchantment. Technically speaking, a partition of an integer n is

---

[18]P. Deligne. *Publ. Maths.,* Institute Hautes E' tudes Sci (IHES), 43, (1974), 273–307.

[19]"Reflections around the Ramanujan centenary", Atle Selberg, reproduced from *Atle Selberg, Collected Papers,* Vol. 1.Springer, 1989, 695–701in *Resonance,* December 1996.

a division of n into any number of positive integral parts. The number of unrestricted partitions of n is denoted by $p(n)$. Ramanujan was also the first mathematician in the world to discover the congruence properties of $p(n)$. The first few examples of $p(n)$ are as follows:

$P(1) = 1$ and $1 = 1$.
$P(2) = 2$ and $2 = 2, 1 + 1$.
$P(3) = 3$ and $3 = 3, 2 + 1, 1 + 1 + 1$.
$P(4) = 5$ and $4 = 4, 3 + 1, 2 + 2, 2 + 1 + 1, 1 + 1 + 1 + 1$.
$P(5) = 7$ and $5 = 5, 4 + 1, 3 + 2, 3 + 1 + 1, 2 + 2 + 1, 2 + 1 + 1 + 1, 1 + 1 + 1 + 1 + 1$.
$P(6) = 11$ and $6 = 6, 5 + 1, 4 + 2, 3 + 3, 4 + 1 + 1, 3 + 2 + 1, 3 + 1 + 1 + 1, 2 + 2 + 2, 2 + 2 + 1 + 1, 2 + 1 + 1 + 1 + 1, 1 + 1 + 1 + 1 + 1 + 1$.

The process continues.

In the case of $p(5)$, it may be noted that $5, 4 + 1, 3 + 2$ are the distinct parts of 5 and $3 + 1 + 1, 2 + 1 + 1 + 1$, and $1 + 1 + 1 + 1 + 1$ are the odd parts of 5. Incidentally, Euler had stated nearly two centuries back that:

The number of partitions of an integer into odd parts equals the number of partitions of that integer into distinct (non-repeating) parts.

This is commonly known as *Euler's theorem*. The example cited above conforms to the celebrated theorem. Euler also developed a useful recurrence relation for partitions by employing his well-known *pentagonal number theorem*.

Thus, with the foundation of Euler and the ingenious methods of Ramanujan, the theory of partitions became a very important area of modern research. The iconic Indian mathematician produced a variety of new and beautiful results. They gave rise to new ideas and opened up new avenues of research in number theory.

In the theory of partitions, Hardy and Ramanujan collaborated and contributed a remarkable theorem. The historical account of this event is quite interesting. In a letter written to G. H. Hardy in 1913, Ramanujan had suggested that there should be an exact formula for the partition function $p(n)$ in terms of continuous functions. Hardy was not convinced about this claim, but thought that it was certainly possible to develop an asymptotic (like an approximate) formula, for $p(n)$ in terms of continuous functions. They together carried out an intricate and cleverly contrived calculation which involved the singularities of the generating function in the unit circle. They were able to successfully obtain an asymptotic formula. They tested the formula and were satisfied to find that $p(200)$ as obtained by the Hardy–Ramanujan formula was identical with that calculated using the Euler's recurrence formula. This result was published as a joint paper by S. Ramanujan and G. H. Hardy and was titled "Une formulae asymptotique pour le nombre des partitions des n".[20]

---

[20] Section 1, (SR. 12).

Another famous number theorist Hans Rademacher subsequently noticed that, simply by replacing the exponential functions with hyperbolic functions, the Hardy–Ramanujan asymptotic series could be converted into a series that converged to $p(n)$. In fact, according to noted number theorist, K. Alladi, in his 1913 letter to Hardy, Ramanujan had used a hyperbolic function to claim an exact formula for a related problem. So, Ramanujan was correct in conjecturing that a similar exact formula could be found for $p(n)$. In the words of K. Alladi[21]:

> … An aspect of Ramanujan's discoveries that comes up time and again is the surprising and unbelievable form of the results. The exact formula for $p(n)$ that Ramanujan conjectured was considered unbelievably good by Hardy who settled for less, namely an asymptotic formula, and Ramanujan agreed (according to Selberg) out of respect for his mentor. In any case, the Hardy–Ramanujan asymptotic formula gave rise to a very powerful analytic method to evaluate the coefficients of series arising in a wide class of problems in additive number theory. The *Circle-method*, originally due to Hardy–Ramanujan and subsequently developed by Hardy–Littlewood and others, is one of the most widely applicable methods today and will continue to be major tool in future.

Professor Atle Selberg, the Fields medallist and one of the greatest among the modern mathematicians, also commented in this regard. He said[22]:

> The paper by Hardy and Ramanujan contained surely a result that was very remarkable in itself; since $p(n)$ is an integer, it allows exact computation on it. But it was not an exact formula, it was a formula with an error tending to zero as n grows and, therefore, $p(n)$ being an integer, one could find the exact value.

> If one looks at Ramanujan's first letter to Hardy, there is a statement there which has some relation to his later work on the partition function, namely about the coefficient of the reciprocal of a certain theta series—a power series with square exponents and alternating signs as coefficients. It gives the leading term in what he claims as an approximate expression for the coefficient. If one looks at that expression, one sees that this is an exact analogue of the leading term in the Radamacher formula for $p(n)$ which shows that Ramanujan, in whatever way he had obtained this, had been led to the correct term of that expression. In the work on partition function, studying the paper, it seems clear to me that it must have been, in a way, Hardy who did not fully trust Ramanujan's insight and intuition, when he chose the other forms of the terms in their expression, for a purely technical reason, which one analyses as not very relevant. I think that if Hardy had trusted Ramanujan more, they should have inevitably ended with the Rademacher series. There is little doubt about that.

In this context, it may be relevant to note that Ramanujan and Hardy in their celebrated paper titled "Asymptotic formula in combinatory analysis"[23] developed the method for deriving an asymptotic formula for the partitions. After Ramanujan's untimely death, Hardy and Littlewood developed the method even further. They also applied it to classical unsolved problems. They were Waring's problem and Goldbach's problem. Focusing on the partitions of a natural number, G. H. Hardy and E. M. Wright wrote[24]:

---

[21] *Ramanujan's Place in the World of Mathematics*. Springer India, 2013.

[22] Talk at TIFR, 1988.

[23] Section 1 (SR. 17).

[24] *Introduction to the Theory of Numbers*, 5th ed. Oxford University Press, London. (1979).

...in spite of the simplicity of the definition of $p(n)$, not very much is known about its arithmetic properties.

This naturally prompts one to discuss Ramanujan's work related to congruence [7].

Firstly, a definition of congruence is necessary to facilitate understanding of the related topic. Suppose a, b are integers and that n is a natural number. By $a \equiv b \pmod{n}$, one means n divides $(b - a)$ and it is said that *a is congruent to b modulo n*. For example, $3 \equiv 24 \pmod{7}$, since $(3 - 24) = -21 = (-3)\,7$, etc.

Referring to the remarks of Hardy and Wright, it may be noted that Ramanujan did make remarkable contributions in this area by publishing his three papers all related to the properties and especially the congruence properties of the partition function $p(n)$. One can refer to the three papers published during 1919–1921:

"Some properties of $p(n)$, the number of partitions of n" [Sect. 2.1, (SR. 20)].

"Congruence properties of partitions" [Sect. 2.1, (SR. 22)].

"Congruence properties of partitions" [Sect. 2.1, (SR. 23)].

Apart from these three papers, the natural genius and ingenuity of Ramanujan combined with the mathematical sophistication of Hardy resulted in two more famous publications mentioned below:

G. H. Hardy and S. Ramanujan. "Asymptotic formulae for the distribution of integers of various types" [Sect. 2.1, (SR. 15)].

G. H. Hardy and S. Ramanujan. "Asymptotic formulae in Combinatory analysis" [Sect. 2.1, (SR. 14)].

The last paper has already been discussed exhaustively. In this connection, another historical development deserves mention. Hardy and Ramanujan wanted to test the correctness of their established formula numerically. At that time, another noted British mathematician P. A. MacMahon had used a recurrence formula of Euler involving the pentagonal numbers and had calculated the first two hundred values of the partition function. So, Hardy thought that using MacMahon's table of values for $p(n)$ would be useful for checking the correctness of the asymptotic formula derived by himself and Ramanujan. But the effect of MacMahon's table on Ramanujan was, to say the least, absolutely dramatic. As soon as he saw the table, he wrote down three congruence properties of $p(n)$. His first congruence stated that:

the number of partitions of an integer of the form $(5m + 4)$ is always a multiple of 5.

The second congruence stated that:

the number of partitions of $(7m + 5)$ is a multiple of 7.

The third congruence stated that:

the number of partitions of $(11m + 6)$ is a multiple of 11.

Ramanujan's statements virtually stunned Hardy. Professor K. Alladi has commented in his book[25]:

> What stunned Hardy was that there are divisibility properties of partitions which are combinatorial objects defined by an additive process...So in principle, anyone staring at the last entry of each column could have observed Ramanujan's first congruence, but one has to be in search of such a property in order to observe it. MacMahon prepared the table and Hardy checked it, but neither of them observed the congruence because they were not looking for such surprising connections! Ramanujan who always had the eye for the unexpected, wrote the congruence down as soon as he saw the table.

It may be noted that even in today's mathematical world, the study of such divisibility properties or congruence properties for partition functions and similar topics attracts many mathematicians and it is still a very active and vibrant area of research in number theory. The role of modular forms, which is an abstract and beautiful area comprising analysis, algebra, and number theory, in the study of such congruence properties is really remarkable.

In this context, it may be noted that Ramanujan had stated in a conjecture that if $m \neq 5, 7, 11$, then in every arithmetic progression there would be infinitely many values of the partition function which would not be multiples of n. Professor Ken Ono of the USA has shown this conjecture to be incorrect. He has proved that for every $m \geq 13$, there are arithmetic progressions with very large moduli, in which the values of $p(n)$ are always multiples of m. Many other famous number theorists are still working in these areas.

Ramanujan's Contributions in Algebra (Hypergeometric Series and Continued Fractions)

In algebra, the two main areas of Ramanujan's work concerns "hypergeometric series" and "continued fractions". According to Prof. G. H. Hardy:

> These subjects suited him exactly, and here was unquestionably one of the greatest masters.

In Ramanujan's "Second Notebook", he has written about hypergeometric series in Chaps. 10 and 11. According to experts in the field, like Hardy, R. P. Agarwal, and others, a substantial part of Ramanujan's work on ordinary hypergeometric series is more or less a rediscovery of the work already done by other mathematicians in the past. Ramanujan himself derived three-term recurrence relations for the hypergeometric series. He obtained his results on continued fractions from these relations. He made asymptotic expansions for hypergeometric series and related functions. His "Second Notebook" contains many such examples. He rediscovered many classical results on hypergeometric series and continued fractions which were already established earlier by Euler, Gauss, and Heine. But, he also discovered many new results on these topics. In his "Lost Notebook", there are many results related to Heine's q-series and some of the q-continued fractions are related to modular functions. Detailed discussions on this type of work of Ramanujan have been done exhaustively by R.

---

[25] *Ramanujan's Place in the World of Mathematics*. Springer India. 2013.

Askey, B. Berndt, G. Watson, and many others. Professor K. G. Ramanathan of TIFR commented:

> The exploitation of the relationship between modular functions and continued fractions is one of the most beautiful aspects of Ramanujan's work on continued fractions.

Rogers–Ramanujan Identities

Ramanujan, while studying an infinite continued fraction, perhaps got motivated to look deeply into identities. The history connected to Roger–Ramanujan identities is indeed rather fascinating. By all available accounts, the identities were first discovered by Rogers in 1894. Rogers was a talented mathematician, but he did not get much recognition for his work. Before leaving for England, while still in Madras, Ramanujan had rediscovered these identities sometime in 1913. However, he had no proof and could not offer any. No one in Hardy's mathematical circle could prove Ramanujan's results. When in 1914, Ramanujan finally arrived in England and settled down in Cambridge, Hardy requested Ramanujan to give the necessary proofs but Ramanujan failed to do so. P. MacMahon had compiled the partition version of the identities and Ramanujan had not contributed anything to that either. In 1915, when the second volume of MacMahon's book, *Combinatory Analysis*, was published, these partition theorems were stated as combinatorial problems, which were still not solved.

Hardy in his book wrote[26]:

> Ramanujan looking through old volumes of the London Mathematical Society came accidentally across Rogers' papers. I can remember very well his surprise and admiration which he expressed for Rogers' work. A correspondence followed in the course of which Rogers was led to a considerable simplification of his original proof.

In 1919, Ramanujan gave a more analytic proof of the identities in his paper titled "Proof of certain identities in combinatory analysis".[27] These beautiful identities appear in Chap. 16 of Ramanujan's second "Notebook".

According to Prof. K. Srinivasa Rao:

> The Rogers–Ramanujan identities arise in the study of Euclidean Lie algebras, and the search for a proof of these identities in this setting led to a deeper understanding of some of the representation theory of these algebras. These identities arise naturally in R. J. Baxter's solution of the hard hexagon model in statistical mechanics.[28]

NOTE: The statement of the first identity is

> The number of partitions of an integer into parts differing by at least 2 equals the number of partitions of that integer into parts which when divided by 5 leave remainder 1 or 4

---

[26] *Ramanujan: Twelve Lectures on Subjects Suggested by his Life and Works.* Chelsea, New York, 1978.

[27] Section 1, (SR. 21).

[28] K. Srinivasa Rao. *Srinivasa Ramanujan*, Revised Edition. East–West Books. Madras, December 2004.

The statement of the second identity is similar. The analytic form in which Ramanujan had communicated the identities to Hardy in a letter in 1913 is quoted below:

> The analytic form of the identities is the equality of two infinite series and two infinite products.

The infinite series represent the generating function of the partitions satisfying difference conditions. The infinite products are the generating functions for partitions satisfying congruence conditions.

Incidentally, it would be appropriate to mention that these two remarkable identities discovered by Ramanujan are clearly related to the first two congruences made by him in partition functions.

Highly Composite Numbers

M. Ram Murty and V. Kumar Murty quote in their book[29]:

> The structural properties of natural numbers and their detailed study form a significant chapter in analytic number theory. In his memoir on highly composite numbers, Ramanujan initiated an important method to study general arithmetic functions. The method has become a dominant theme in current research. Surprisingly, these studies take us into allied areas of transcendental number theory and an intimate study of the Riemann zeta function.

Before proceeding any further, it would be necessary to give some definitions.

Composite Number

A natural number n is said to be *composite* if it has a divisor different from 1 and itself. Next, Ramanujan's question was if n is a composite number then what makes it highly composite? He gave the following definition for "highly composite numbers".

Highly Composite Number

A natural number n is a *highly composite number* if $d(m) < d(n)$ for all $m < n$, where $d(r)$ denotes the number of distinct positive divisors of r.

If one considers the primes and composite numbers in Z, 1 is a unit element and 2 is a prime. But according to Ramanujan's definition, both of them become highly composite numbers.

In 1915, Ramanujan published his paper titled "Highly Composite Numbers".[30] In this paper, he gave a complete classification of highly composite and superior highly composite numbers. Ramanujan in the introduction to the paper wrote:

> The number $d(N)$ of divisors of N varies with extreme irregularity as N tends to infinity, tending itself to infinity or remaining small according to the form of N. In this paper, I prove a large number of results which add a good deal to the knowledge of the behaviour of $d(N)$.

In the concluding part of the introduction, Ramanujan wrote:

---

[29] "Prime numbers and highly composite numbers. "In *The Mathematical Legacy of Srinivasa Ramanujan*. Springer India, 2013.

[30] Section 1, (SR. 8).

I prove also that two successive highly composite numbers are asymptotically equal, that is, that the ratio of two consecutive such numbers tends to infinity. These are the most striking results. More precise one will be found in the body of the paper. These results give us a fairly accurate idea of the structure of a highly composite number. I then select from the general aggregate of highly composite numbers a special set which I call 'superior highly composite numbers'. I determine completely the general form of all such numbers, and I shew how a combination of the idea of a superior highly composite number with the assumption of the truth of the Riemann's hypothesis concerning the roots of the ξ-function leads to the even more precise results concerning the maximum order of $d(n)$. These results naturally differ from all which precede in that they depend on the truths of a hitherto unproved hypothesis.

Ramanujan had calculated the first 102 composite numbers. The first few superior composite numbers selected by him are 2, 6, 12, 60, 120, 360, 2520, ..., etc. On the strength of his paper, Ramanujan was awarded the B. A. degree by research of the Cambridge University. It may further be noted that in the paper under discussion, Ramanujan studied $Q(x)$, the number of highly composite numbers $\leq x$, in great detail. He devoted an entire section to this investigation. He also studied related arithmetical functions $d_k(n)$ counting the number of ways of writing n as a product of k numbers. He also investigated the generalized divisor sum functions $\sigma_s(n)$, the sum of the sth powers of the positive divisors of n. Strangely enough, in the final publication of the London Mathematical Society, sections 53 to 75 of his paper were not printed, because of the financial crunch of the society.

Fortunately, the unpublished sections were written in the *Lost Notebook*. Otherwise, they would have been lost to the world. His three notable theorems stated in the paper are widely known. They are

Theorem: There is an Infinite Number of Highly Composite Numbers.
Theorem: Successive Highly Composite Numbers Are Asymptotically Equivalent.
Theorem: If $Q.(x)$ is the number of highly composite numbers $\leq x$, then $\lim Q(x)/\log x = +\infty$.

The determination of $Q(x)$ was of great interest to Ramanujan.

Diophantine Equations
Certain algebraic equations in number theory, which require solutions in integers, are called *Diophantine equations*. In his *Second Notebook*, Ramanujan had given solutions of Diophantine equations:

$$X^3 + Y^3 + Z^3 = U^3 \text{(Euler's equation)}$$

in rational numbers. The solutions were given as

$$X = m^7 - 3m^4(1+p) + m(2+6p+3p^2),$$
$$Y = 2m^6 - 3m^3(1+2p) + 1 + 3p + 3p^2,$$
$$Z = m^6 - 1 - 3p - 3p^2,$$
$$U = m^7 - 3m^4 p + m(3p^2 - 1),$$

where m and p are arbitrary numbers.

In this connection, it would be interesting to recall Hardy's comments. He wrote about Ramanujan:

> He could remember the idiosyncrasies of numbers in an almost uncanny way. It was Little-wood who said that every positive integer was one of Ramanujan's personal friends. I remember going to see him once when he was lying ill in Putney. I had ridden in taxi-cab number 1729, and remarked that the number seemed to me a rather dull one, and I hoped that it was not an unfortunate omen. "No", he replied. "It is a very interesting number; it is the smallest number expressible as a sum of two cubes in two different ways." I asked him, naturally, whether he could tell me the solution of the corresponding problem for fourth powers; and he replied, after a moment's thought, that he knew no obvious example, and supposed that the first such number must be very large.

Actually, $1729 = 1^3 + 12^3 = 9^3 + 10^3$. In fact, however, the simplest known solution of $X^4 + Y^4 = Z^4 + t^4$ has been given by Euler as $59^4 + 158^4 = 133^4 + 134^4 = 635,318,657$. But Euler was not able to give any general solution. In his "Notebook", Ramanujan had shown that if $\alpha^2 + \alpha\beta + \beta^2 = 3\lambda\gamma^2$, then $(\alpha + \lambda^2\gamma)^3 + (\alpha\beta + \gamma)^3 = (\lambda\alpha + \gamma)^3 + (\beta + \lambda^2\gamma)^3$. It may be noted that in the above 4-parameter solution for Euler's equation, if one substitutes $\alpha = 3$, $\beta = 0$, $\lambda = 3$, and $\gamma = 1$, then one will get $12^3 + 1^3 = 10^3 + 9^3 = 1729$. Ramanujan's natural genius is reflected in the spontaneity of his answer.

Probabilistic Number Theory

During his stay in England, Ramanujan in collaboration with Prof. Hardy wrote six research papers. One of these papers was titled "The normal number of prime factors of a number n".[31]

What does the sentence "the number of prime factors of an integer n" mean? It may actually be defined in two ways. Firstly, as the number of "distinct" prime factors in the prime factorization of n. This is symbolically defined by $\Omega(n)$. For example, $\Omega(18) = \Omega(3^2, 2) = 2$. Secondly, if it is defined as the total number of prime factors, it would mean the number of distinct prime factors counted with multiplicity, symbolically denoted by $\Omega(n)$, $\Omega(n) = \Omega(3^2. 2) = 3$. In this paper, Hardy and Ramanujan have extensively studied both these functions.

Hardy and Ramanujan also explored the properties of "round numbers". *Round numbers* are numbers which have substantially more prime factors. It may be mentioned here that prime numbers have been studied from very ancient times in Greece as well as in India. But Ramanujan and Hardy were the first mathematicians to carry out systematic studies on the number of prime factors among integers. One of the significant results that Hardy and Ramanujan showed in the abovementioned paper was that almost all integers n have log log n prime factors.

In the said paper, Hardy and Ramanujan for the first time introduced the concept of "normal order" of an arithmetical function. Divisor functions do not possess a normal order. But if $\omega(n)$ is defined as the number of prime factors of n, then $\omega(n)$

---

[31] Section 1, (SR. 16).

has a normal order. Actually, in the paper mentioned above Hardy and Ramanujan showed that the normal order of $\omega(n)$ was log log n.

It may be noted that in this particular paper, the concepts involved such as "normal order" and "almost all integers" clearly display a probabilistic flavor. That is why it is credited for opening up a new branch of number theory namely "probabilistic number theory". This theorem is regarded as the first result in this new discipline.

But the true significance of this paper came to light many years after its publication. In 1934, Hungarian mathematician Paul Turán gave a nice and simple proof of the problem. This proof indicated the presence of important probabilistic principles at the core of the Hardy–Ramanujan result. This led to the discovery of a new branch of mathematics known as "probabilistic number theory". How the event of this discovery took place in Princeton, USA, has been beautifully described by K. Alladi in his book[32]:

> In 1939 during a lecture in Princeton, the great mathematician Mark Kac described some possible new applications of Probability Theory, in particular to problems related to the Hardy–Ramanujan results on the number of prime factors. But in order to carry out these ideas of Kac, various auxiliary results from Number Theory were required. Fortunately, Erdös was in the audience. As soon as Kac finished his lecture, Erdös walked up to him and showed how "sieve methods" from Number Theory could be used to complete the argument of Kac. Thus the celebrated Erdös–Kac theorem was proved, and with this 'probabilistic number theory' was born.

Probabilistic number theory is a very vibrant area of research in modern times. Peter Elliott, in his two-volume book, has made detailed discussions on the many major advances in this area starting with the work of Hardy–Ramanujan. P. Erdös and M. Kac, P. D. T. A. Elliot, A. Selberg, C. Spiro, and many others have made valuable contributions in this and related areas.

Mock Theta Functions

From his deathbed in Madras, Ramanujan thought and wrote about his last discovery to Prof. Hardy and named it "Mock-theta functions". After his sad and untimely demise in 1920 and particularly after the discovery of his *Lost Notebook*, many eminent mathematicians like G. N. Watson, A. Selberg, and G. E. Andrews studied the examples of this last discovery and gave rigorous proofs of some of the statements he had made. In 1988, D. R. Hickerson, in his two papers titled "A proof of the mock theta conjectures"[33] and "On the seventh order mock theta function",[34] proved many of Ramanujan's conjectures related to mock theta functions. However, no unifying theory emerged from these papers. Only in 2002, when S. Zwegers completed his doctoral thesis under D. Zagier that one finds there an intrinsic characterization of mock theta functions. Later, Bringmann and Ono used this theory to resolve several open problems in Combinatorics and the theory of q-series.

---

[32] *Ramanujan's Place in the World of Mathematics*. Springer, India, 2013.
[33] *Invent. Math.*, 94, (1988), 639–660.
[34] *Invent. Math.*, 94, (1988), 661–677.

D. Zagier wrote in "Ramanujan's mock theta functions and their applications"; (d'apre's Zwegers and Bringmann-Ono), Sem. Bourbaki 69, (2007)] in his lucid survey of Ramanujan's work:

> Ramanujan used the word 'theta function' where we would say 'modular form' today so that 'mock-theta functions' meant something like 'mock-modular forms'. More precisely, a mock-theta function is a mock-modular form of weight 1/2. All of the 17 examples of Ramanujan are of the form,
>
> $$\frac{A_{n+1}(q)}{A_n(q)} = R(q, q^n)$$
>
> where $A_n(q)$ is in $Q(q)$, for all $n \geq 1$, and for some fixed rational function.
>
> Such series are called *q-hypergeometric series*.[35]

In 2001, Zwegers in his doctoral thesis discovered the relation between non-holomorphic modular forms, Lerch sums, and indefinite theta series. He established that mock theta functions were actually the "holomorphic parts" of real analytic modular forms of weight ½. This last discovery of Ramanujan is still a very vibrant and challenging area of research in number theory. Work in this direction is both national and international. According to noted number theorist K. Alladi:

> There are many who consider the mock-theta functions are among Ramanujan's deepest contributions. (They were discovered by Ramanujan in India after his return from England a few months before his death. In his last letter to Hardy, Ramanujan states that he made a very significant discover, namely the mock theta functions of order 3 or 5)…Some consider the theta functions as the greatest mathematical discovery of the nineteenth century. Theta functions are extremely interesting because they satisfy transformation properties. The mock-theta functions are like the theta functions in the sense that the circle method can be used just as effectively to calculate their coefficients, but they do not satisfy exact transformation formulae. The example, which Ramanujan sent to Hardy, was a well-known theta-type series but with mild changes in sign which made them mock-theta functions.

Srinivasa Ramanujan was the fountainhead of research in the area of number theory in India. He influenced generations of number theorists and his questions and conjectures are still awe-inspiring. As noted number theorist Prof. K. Alladi commented:

> Ramanujan lives even today through the many questions he has raised. Deligne's winning the Fields Medal by solving the Ramanujan hypothesis is a monumental inscription to his illustrious and ever-lasting memory.

The renowned Astrophysicist and Nobel Laureate Prof. S. Chandrasekhar had very rightly said:

> As long as people do mathematics, the work of Ramanujan will continue to be appreciated.

---

[35]M. Ram Murty and V. Kumar Murty. *The Mathematical Legacy of Srinivasa Ramanujan*. Springer, India, 2013.

Annexure

[1] *q-series*: Mathematically speaking, a q-series has expressions of the type $(a; q)_n$ in the summands. However, many q-series do not have such products in their summands, but they may arise as limiting cases of series containing products of the type

$$(a)_0 := (a; q)_0 := 1, (a)_n := (a; q)_n, \text{ where } (q)_n = (q; q)_n = \prod_{k=1}^{n} (1 - q^k).$$

[2] *Theta functions*: The functions which are intimately connected with q-series in their theories are known as *theta functions*.

[3] *Elliptic functions*: In complex analysis, an *elliptic function* is a meromorphic function that is periodic in two directions. A periodic function of a real variable is defined by its value on an interval. An elliptic function is determined by its value on a fundamental parallelogram, which then repeats in a lattice.

[4] *Meromorphic functions*: A *meromorphic function* is a single-valued function that is analytic in all but possibly a discrete subset of its domain and at those singularities, it must go infinity like a polynomial (i.e., these exceptional points must be poles and not singularities).

[5] *Hypergeometric series*: A series x_n is called *hypergeometric* if the ratio of successive terms x_{n+1}/x_n is a rational function of n. If the ratio of successive terms is a rational function of q^n, then the series is called a *basic hypergeometric series*. The number q is called the *base*.

[6] *Modular functions*: In mathematics, a modular form is a complex analytic function on the upper-half plane satisfying a certain kind of functional equation with respect to the group action of the modular group, and also satisfying a growth equation. *Modular functions* are special cases of modular forms, but not vice versa.

[7] *Congruence*: Suppose a, b are integers and that n is a natural number. By writing $a \equiv b \pmod{n}$ one means n divides $(b - a)$, and it is said that *a is congruent to b modulo n*. For example, $3 \equiv 24 \pmod{7}$, since $3 - 24 = -21 = (-3) 7$, etc.

2.1.2   K. Ananda Rau (1893–1966)

Next, we look into the contributions in the development of the school of number theory made by K. Ananda Rau. According to a famous number theorist of present-day India, R. Balasubramanian:

> If Ramanujan's influence in India is the tree…then K. Ananda Rau is the root.

K. Ananda Rau (Fig. 2.2) was almost a contemporary of S. Ramanujan. He was born in a well-to-do and well-connected family in erstwhile Madras (present-day Chennai) in 1893. Unlike Ramanujan, he had a solid foundation of formal education. He completed his schooling from the reputed Hindu High School and then pursued his college education from the famous Presidency College of the University of Madras.

Fig. 2.2 K. Ananda Rau
(1893–1966)

Both the institutions that he attended were among the best in India at that time. Throughout his student days, he had an excellent academic career.

In 1914, only a few months after Ramanujan, K. Ananda Rau sailed to England and took admission to the King's College of the University of Cambridge. In 1916, he completed the Mathematics Tripos examination with a first-class honors. He was subsequently elected a Fellow of the King's College. Shortly after that he met Prof. G. H. Hardy and was encouraged by him to take up research in mathematics. Influenced, and to some extent guided, by Prof. Hardy, K. Ananda Rau worked mainly on mathematical analysis. But mention must be made of the paper entitled "The infinite product for $(s - 1)\zeta(s)$",[36] where he gave a proof of factorization of $(s - 1)\zeta(s)$ by using Cauchy's theorem of residues [8], similar to some extent like the factorization of $\sin z$. K. Ananda Rau also pointed out that his method of proof could be used to factorize the function $L(s)$ of analytic number theory. Apart from this paper, around 1929, he became interested in number theory per se, and in a series of three research papers mentioned below, he studied the behavior of elliptic modular and elliptic theta functions. The papers are:

- "On the boundary behaviour of elliptic modular functions" [Sect. 1, (KAR. 2)]
- "Additional note on the boundary behaviour of elliptic modular functions" [Sect. 2.1, (KAR. 3)]
- "On the behaviour of elliptic theta functions near the line of singularities" [Sect. 2.1, (KAR. 4)].

Ananda Rau studied the boundary behavior of the elliptic modular function:

---

[36]Section 1, (KAR. 1).

$$\Theta_3(0|\tau) = 1 + \sum_{n=0}^{\infty} q^{n^2}; \text{ where } q = e^{i\pi\tau}, \tau = \xi + iy, y > 0.$$

Earlier, Hardy and Littlewood had proved that if ξ is irrational and its continued-fraction expansion has bounded partial quotients, then there are positive constants K_1, K_2 for which a certain inequality existed. In the third paper mentioned above (published in 1933), Ananda Rau proved the converse result and elaborated it by showing that the partial quotients of ξ can be divided into two sets such that the boundedness of each set implied one half of the Hardy–Littlewood inequality that is involving either K_1 alone or K_2 alone and conversely.

In the first paper mentioned above, Ananda Rau showed that the limit values arranged themselves on a finite number of well-defined curves which could be specified in terms of the continued fraction for ξ. In the second paper of the series, he extended the results obtained in the first paper to certain more general types of irrational ξ (including numbers like e and e^2) whose continued-fraction expressions exhibited some kind of periodicity, in particular, residue class periodicity (mod 8).

Much later in the fifties of the twentieth century, K. Ananda Rau again started taking fresh interest in number theory. At that time, he tried to represent numbers as sums of an even number of squares. His paper entitled "On the representation of a number as the sum of an even number of sequences" was published in the journal of Madras University.[37] Apart from this paper, he published four other papers on number theory, all related to quadratic forms [9], as mentioned below:

- "On the summation of singular series associated with certain quadratic forms (I)" [Sect. 2.1, (KAR. 6)]
- "Applications of modular equations to certain quadratic forms" [Sect. 2.1, (KAR. 7)]
- "Relation between sums of singular series associated with certain quadratic forms" [Sect. 2.1, (KAR. 8)]
- "On the summation of singular series associated with certain quadratic forms (II)" [Sect. 2.1, (KAR. 9)].

Of these four papers, the paper which was published in the jubilee issue of the *Journal of the Indian Mathematical Society* in 1960 has been considered to be of significant importance in the area of number theory. K. G. Ramanathan, a noted number theorist of India, gave a resume for this paper. C. T. Rajagopal in his article on "K. Ananda Rau" elaborated on this resume and wrote[38]:

Let r(n) denote the number of representations on n by the quadratic form $X_1{}^2 + X_2{}^2 + C(X_3{}^2 + X_4{}^2)$, where C > 0 is an integer. Let $q = e^{i\pi\tau}$, $\tau = X + iY$, Y > 0 and according to Hecke, $\Theta^2(q)\, \Theta^2(q^c) = W(q) + Z(q)$, where $W(q^2)$ is an Eisenstein series and $Z(q^2)$ is a cusp form.

---

[37] Section 1, (KAR. 5).

[38] *Journal of London Math. Soc.*, 44, (1969), 1–6.

In the paper under discussion, he has obtained explicit expressions for $W(q)$ and $Z(q)$, when $C = 3, 5, 6, 7, 11$, or 23 by using modular equations of degree C or $C/2$ accordingly as C is an odd prime or twice an odd prime. For $W(q)$, the expression that was obtained by him was in terms of the Lambert-type series. The expressions for $C = 3, 5, 6, 7$ had already been obtained by Kloosterman. But for $C = 11, 23$, the results obtained by K. Ananda Rau were new.

K. Ananda Rau was an accomplished mathematician and was himself a good researcher. But his contributions in teaching and guiding students and research scholars were extraordinary. Ananda Rau returned from Cambridge, England, in 1919. Shortly after that, he was appointed a professor of mathematics at his own *alma mater,* Madras Presidency College, at the young age of 26. Noted mathematician V. Ganapathy Iyer was one of his students. He commented:

> As a student I used to feel that his exposition of any topic was so clear and impressive that I need not study the topic again.

C. T. Rajagopal, a doctoral student of Prof. Ananda Rau, said that Ananda Rau's way of working with research scholars was rather novel. He encouraged them and expected them to formulate their own problems. Later, he would discuss the problems with them. In this way, he brought out the best in them. Special emphasis is being given to this quality of Ananda Rau, because many of the stalwarts of the number theory school of South India were mentored by him. Starting from T. Vijayaraghavan, S. S. Pillai, K. Chandrasekharan to C. T. Rajagopal, all were his students. All of them made rich contributions in developing the school of research on number theory in India.

Annexure

[8] *Cauchy's theorem of residues*: In complex analysis, Cauchy's residue theorem is a powerful tool for the evaluation of line integrals of analytic functions over closed curves. It is often used to compute real integrals as well.

[9] *Quadratic forms*: It is a homogeneous polynomial of degree two in a number of variables.

In the thirties of the twentieth century, two students of Prof. K. Ananda Rau made very notable contributions to developing the school of research on number theory in India; they were S. S. Pillai and T. Vijayaraghavan. Both of them hailed from the southern part of India. But for some years, both had shifted to the pre-Independent undivided Bengal. So apart from their contributions to South Indian school of research on number theory, they were also responsible for initiating number theory-based research in Bengal.

2.1.3   Subbayya Sivasankaranarayana Pillai (1901–1950)

Subbayya Sivasankaranarayana Pillai (or S. S. Pillai) was born near Courtallam in the Tirunelveli district of erstwhile Madras Presidency in 1901 (Fig. 2.3). He lost his mother in infancy and his father passed away when he was in the matriculation class. He was a talented student right from his school days but had to struggle hard with persistent poverty.

With financial help from one of his school teachers, whom Pillai always addressed as *sastriar*, and a scholarship which he obtained for his performance in the matriculation examination, he managed to complete his B.A. from Maharajah's College at Trivandrum (now, Thiruvananthapuram). Around 1927, he moved to Madras (now, Chennai). He received a stipend from the University of Madras, which enabled him to carry on the postgraduate studies. It was his first exposure to higher mathematics. In the University of Madras, he came in contact with the noted mathematician Prof. T. Vaidyanathaswamy (1884–1960). Pillai started doing research in mathematics under the supervision of Prof. K. Ananda Rau. As already discussed, Ananda Rau himself had made important contributions to number theory. So, S. S. Pillai in all probability was influenced by him and decided to investigate problems related to numbers, Euler's totient functions, etc. Before elaborating on his research work, it would be relevant to quote the comments of several noted number theorists of India, so that an idea can be made about the importance of his contributions. Professor R. Balasubramanian and R. Thangadurai observed that:

> S. Sivasankaranarayana Pillai (1901–50) is probably best known to the mathematical community for the Waring's problem and exponential Diophantine equations. He has influenced

Fig. 2.3 S.S. Pillai
(1901–1950)

number theory with his valuable contributions as Pillai's theorems, questions and conjectures. They have led the way to a considerable research.[39]

Another noted number theorist Prof. K. Chandrasekharan wrote:

It is perhaps fitting that something ought to be said about his contribution to the solution of Waring's problem, which is certainly his best piece of work, and one of the very best achievements in Indian mathematics since Ramanujan.[40]

Now coming to his research career, it may be noted that from 1927 to 1929, Pillai published almost 7 research papers. The first few problems were suggested to him by K. Ananda Rau or R. Vaidyanathaswamy. During this period, Pillai carried out some studies of arithmetic functions, $\varphi(n)$, and on the representation of numbers, etc. Apart from his supervisor K. Ananda Rau, Ramanujan must have been the indirect influence, which prompted Pillai to work in that direction of number theory. From his published papers, three made the contents of his M.Sc. dissertation. After obtaining his M.Sc. degree in 1929, he was selected as a lecturer in the mathematics department of Annamalai University at Chidambaram in Madras Presidency. Pillai took up the job and worked there until 1941. Those were S. S. Pillai's most productive years. Between 1930 and 1940, in 10 years, he had published as many as 50 research papers. His most remarkable work on Waring's problem [10] was carried out during this period. His 20-year long correspondence with S. Chowla, another outstanding number theorist of India, resulted in 5 publications during 1930–1940. During his tenure at Annamalai University, he submitted his D.Sc. thesis and was awarded the degree by the University of Madras. He was the first mathematician to obtain a doctorate degree in mathematics from Madras University. S. S. Pillai worked on various areas of number theory and in this context, K. Chandrasekharan observed:

The volume of his research is therefore quite large; what is even more remarkable about it is that 'it is fresh and has a distinct touch of originality'. Pillai was a lover of problems, and of genuinely difficult problems. The theory of numbers has always had an abundant supply of problems, and so he was never forced to expand his interests. This resulted in a concentration of efforts that is well typified in his mastery of what may be called 'asymptotic analysis'.

The discussions on Pillai's major research contributions in his chosen field of number theory will naturally start with Waring's problem.

In 1640, Pierre de Fermat (1607–1665) conjectured among other results that every number is the sum of 4 squares. But he did not prove it. 130 years after Fermat stated it, in 1770 Lagrange proved and published the result. Suppose one intends to express n as a sum of squares, say, in the form $n = x_1^2 + x_2^2 + \cdots + x_s^2$. However, Fermat's theorem implies that it is possible to find s independent of n, and its value is 4. Waring's problem is actually the generalization of this result for kth powers. In 1770, Edward Waring (1732–1798), Lucasian Professor of mathematics at

---

[39]"Collected works of S. Sivasankaranarayana Pillai," Vol. 1.In *Ramanujan Math. Soc.*, (2010), R. Balasubramanian and R. Thangadurai (Eds.).

[40]"S. S. Pillai". In *Journ.Indian Math. Soc.* (N. S.), Part A, 15, (1951), 1–10.

Cambridge University, conjectured in his "Meditations algebraic", that every number is the sum of 4 squares, 9 cubes, 19 biquadrates, and so on. Now let $g(k)$ be the least value of s, such that every number is the sum of s, kth powers. Or in other words, the equation $n = x_1{}^k + x_2{}^k + \cdots + x_s{}^k$ is solvable for every value of n. It is not clear whether the least value of s will increase with n or not. Waring's conjecture only implies that $g(k)$ is independent of n and that $g(2) = 4$, $g(3) = 9$, $g(4) = 19$, and so on. But Waring did not give any proofs for his stated results. Historically speaking, the earliest known work on Waring's problem was carried out in 1772 by J. A. Euler (1734–1800), son of legendary mathematician Leonhard Euler (1707–1783). He worked out and showed that for all $k \geq 1$, the lower bound $g(k) \geq 1 + 2^k - 2$, where $1 = (3/2)^k$.

The next important step was taken in 1858 when J. Liouville (1809–1882) proved that $g(4) \leq 53$. In 1903, a lesser known number theorist A. Wieferich showed that $g(3) = 9$. In 1912, Kampner (1880–1973) proved the same. In fact, for values of k different from 1 and 2, these were the first determination of $g(k)$. E. Landau (1877–1938) showed that every large number is the sum of 8 cubes. D. Hilbert (1862–1943) was the first mathematician who succeeded in proving the existence of $g(k)$. In the context of Hilbert's proof, Prof. G. H. Hardy remarked:

> Within the limits which it has set for itself, it is absolutely and triumphantly successful, and it stands with the work of Hadamard and de la Vallee Poussin in the theory of primes as one of the landmarks in the modern history of theory of numbers.

Landau's result had special importance. Because of that result, Hardy and Littlewood introduced another more fundamental function $G(k)$. The meaning of Landau's "singularly beautiful theorem" is that "the number of numbers which require cubes is finite". In fact, Dickson later proved that 23 and 239 are the only two numbers which require as many as 9 cubes. $G(k)$, the function introduced by Hardy and Littlewood, was defined as the *least value of s*, such that every sufficiently large number from a certain integer onwards is the sum of s, kth powers. In the light of this definition, Fermat's result would mean $g(2) = G(2) = 4$ and Landau's result becomes $G(3) \leq 8$.

Now getting back to S. S. Pillai, it would be pertinent to mention that he seriously took up research work in this area only in the latter part of the 1930s. In connection with his research on Waring's problem, Pillai himself wrote:

> The original form of the Waring's problem has been completely solved independently by Dickson and me. Strictly speaking, I have got the priority.

Again in his own assessment of his research work on Waring's problem, S. S. Pillai nicely described the developments of research in that area post Hilbert's proof. He wrote:

> Hilbert's proof was a pure existence proof. The next advance was due to Hardy and Littlewood. In the year from 1919 to 1927, they attacked the problem in a series of memoirs. They discovered a method of applying Cauchy's theorem to the problem. It may be remarked that the Farey-dissection of the circle, which plays a fundamental role in all subsequent investigations, first appear in Hardy–Ramanujan's memoir on partitions. By their important method,

besides finding upper bounds for $G(k)$, Hardy and Littlewood were able to find an asymptotic formula for the number of representations of a number as the sum of s, k-th powers, when s is large.

After a gap of several years, I. M. Vinogradov (1891–1993) in 1928 published a paper titled "On Waring's theorem",[41] in which he simplified the method of Hardy and Littlewood and was able to recover the bound in a beautiful, short, and simple way. In 1935, Vinogradov made further progress and was able to prove the following beautiful result:

$$\mathrm{Lim} \frac{G(k)}{K \log k} \leq 6$$

In this context, noted number theorist Prof. R. Balasubramanian has remarked:

With the field so set by Vinogradov's method, the years 1935 to 1940 saw an intense amount of work on the determination of $g(k)$. A leading role was played by S. S. Pillai, and essentially independently by L. E. Dickson.

In September 1935, Pillai proved that

$$g(k) = 2^k + \left(\frac{3}{2}\right)^k + \mathrm{O}\left(\frac{4}{3}\right)^k$$

The result was proved without applying Dickson's method of ascent. S. S. Pillai was perhaps the first number theorist who noticed that a condition on the Diophantine approximation of $(3/2)^n$ was necessary to get $g(k) = B$, where

$$B = 2^k - \left(\frac{3}{2}\right)^k - 2$$

In 1936, Pillai gave a remarkable and practically a complete solution for the determination of $g(k)$, for $k > 7$. Ultimately, S. S. Pillai showed that $g(7) = 143$ and in a later paper in 1940, he proved that $g(6) = 73$ [$g(6) = 2^6 - (3/2)^6 - 2 = 73$ approx.]. In 1964, Jing-Run Chen proved that $g(5) = 2^5 + (3/2)^5 - 2 = 37$, and in 1986, R. Balasubramanian, J. M. Deshouillers and F. Dress proved that $g(4) = 2^4 - (3/2)^4 - 2 = 19$.

Pillai's interest in Waring's problem was not limited to the determination of $g(k)$. He also contributed to Waring's problem with the powers of primes and to the problem with indices $\geq n$. He alone published a total of 16 original research papers on Waring's problem and related areas. He also wrote some notes on these topics and they still remain unpublished.

The next area of theory of numbers, where Pillai has made notable contributions, is "exponential Diophantine equations" [10]. S. S. Pillai himself has stated:

---

[41] Izvestiya Akademii Nauk, SSSR, No. 4, (1928), 393–400.

Apart from an implied result of Polya, I am the first to consider systematically the equation
of the type $a^x - b^y = c$.

As already mentioned during the discussions on Waring's problem, Pillai was perhaps
the first mathematician to show deep connections with Diophantine problems in his
work on Waring's problem. As early as in 1930, he published his first paper on
Diophantine equations "On some Diophantine equations".[42] Later, in his paper titled
"On the inequality $0 < a^x - b^y < n$",[43] Pillai had proved that for any fixed positive
integers a and b, both at least 2, the number of solutions (x, y) of the Diophantine
inequalities $0 < a^x - b^y \leq c$ is asymptotically equal to

$$\frac{(\log c)^2}{2(\log a)(\log b)}; \text{asc} \to \infty$$

At that time, many number theorists the world over were working on similar and
related problems. P. Ribenboim in his book titled *Catalan's Conjecture*[44] has made
a chronological survey of the study of such problems. He has specially referred to
the works of C. Stormer (1908), A. Thue (1908), G. Polya (1918), T. Nagell (1925
and 1945), and S. S. Pillai (1945). In 1932, Pillai went on to prove that for a tending
to plus infinity, the number $N(a)$ of solutions (x, y), both positive integers > 1, with
$0 < x^y - y^x \leq a$, will satisfy

$$N(a) \sim \frac{1}{2} \frac{(\log a)^2}{(\log \log a)^2}$$

The work on Diophantine equations started by S. S. Pillai in 1931 was pursued by
A. Herschfeld during 1935–1936. In two papers published by him in the Bulletin of
the American Mathematical Society, Herschfeld showed that if c is an integer with
sufficiently large $|c|$, then the Eq. $2x - 3y = c$ has at most one solution (x, y) in
positive integers x and y. However, in his two papers namely,

- "On $a^x - b^Y = c$" [Sect. 1, (SSP. 28)]
- "A correction to the paper 'On $a^x - b^Y = c$'" [Sect. 2.1, (SSP. 29)],

Pillai extended Herschfeld's results to the more general exponential Diophantine
equation $a^x - b^y = c$, where a, b, and c are fixed nonzero integers with gcd $(a, b) =$
1 and $a > b \geq 1$. To prove this result, Pillai made use of Siegel's theorem.[45] Finally,
Pillai was able to deduce the number of integers in the range $[1, n]$, which can be
expressed in the form $a^x - b^y$ and is asymptotically

---

[42] Section 1, (SSP. 13).

[43] Section 1, (SSP. 14).

[44] Academic Press, Boston, MA, 1994.

[45] C. L. Siegel: Abhandlungen Akad, Berlin (1929), No. 1, 705.

$$\frac{(\log n)^2}{2(\log a \log b)}; \text{ as} n \to \infty$$

In his paper titled "On numbers of the form $2^a 3^b$, I",[46] Pillai elaborated on a result by S. Ramanujan. Ramanujan had claimed that the number of integers of the form $2^a \bullet 3^b$, which are bounded by a given number x, is approximately

$$\frac{\log(2x)\log(3x)}{2(\log 2 \log 3)}$$

In collaboration with A. George, this study was further pursued and resulted in the publication of the paper titled *"On numbers of the form $2^a 3^b$, II".*[47]

In this connection, a comment by number theorist Michel Waldschmidt is of historical importance. He wrote in the abstract of his paper titled "Perfect powers: Pillai's works and their developments"[48]:

A perfect power is a positive integer of the form a^x, where $a \geq 1$ and $x \geq 2$ are rational integers. Subbayya Sivasankaranarayana Pillai wrote several papers on these numbers. In 1936, and again in 1945, he suggested that for any given $k \geq 1$, the number of positive integer solutions (a, b, x, y), with $x \geq 2$ and $y \geq 2$, to the Diophantine equation $a^x - b^y = k$ is finite.

This suggestion known as *Pillai's conjecture* led to many open problems. A lot of research has been carried out by many mathematicians in this direction, particularly in the international arena. In India, Prof. R. Balasubramanian and his students have done some work on related areas.

In the literature of number theory, several classes of exponential Diophantine equations occur, which are of the form $Ax^p + By^q = Cz^n$, where A, B, and C are fixed positive integers, while x, y, z, p, q, and r are non-negative integers. Again some of these integers may be taken to be fixed or composed of fixed primes. A subclass of such equations, where $q = 0$ (and if $y = 1$) is called *Pillai's Diophantine equation*.

Some of Pillai's important results on exponential Diophantine equations are mentioned in J. F. Koksma's book titled *Diophantische Approximationen*.[49] S. S. Pillai has published 14 papers on Diophantine equations and 2 on perfect powers. Later, many mathematicians like T. N. Shorey, R. Balasubramanian, P. P. Prakash of India, and R. Tijdeman, A. J. Van der Poorten, A. Schinzel, etc., from other countries, have done extensive research in these areas. Most of S. S. Pillai's work on Waring's problem and a substantial part of his work on Diophantine equations were done by him during his engagement at the mathematics department of Annamalai University.

Another very important and novel event took place during this period. From the late 1920s, S. S. Pillai came in touch with another outstanding number theorist of India. He was Sarvadaman Chowla (S. Chowla) who hailed from Lahore in undivided

---

[46]Section 1, (SSP. 59).

[47]Section 1, (SSP. 60).

[48]*Collected Works of S. S. Pillai*. Ramanujan Mathematical Society, India.

[49]*Ergebnisse der Math.*, IV, 4, Berlin, 1936.

Panjab. He was an extremely talented and internationally acclaimed mathematician. He made remarkable contributions to number theory and was the pioneer researcher who set up the Panjab school of research on number theory. For his pathbreaking research on the discipline and allied areas, he has been named "an ambassador of number theory". Detailed discussions on Chowla and his contributions would be done later. But for the present, attention would be focused on the 20-year long correspondence which took place between S. S. Pillai and S. Chowla.

This remarkable and prolonged correspondence resulted in the joint publication of 3 valuable research papers on "Omega results" [11] and 2 papers on other topics of the theory of numbers. According to Prof. R. Balasubramanian:

> Starting in the late 1920s, and up to one month before his death, S. S. Pillai and S. Chowla maintained a regular correspondence for about 20 years. Through the years that they corresponded, Chowla and Pillai published 5 joint papers and met numerous times at the conventions of the Indian Mathematical Society, of which both were active members. Their correspondence is very revealing for the insight it gives us, as one may expect, into the mathematics and friendship of Chowla and Pillai.

In the earliest letter from Chowla dated January 3, 1929, among various mathematical discussions, it is also mentioned that 175,95,9000 is the least number that can be expressed as a sum of two cubes in three different ways. Chowla ended the letter with the sentence:

> I hope that we will soon begin proper work.

After the exchanges of quite a few more letters, Pillai and Chowla published their first joint paper titled "On the error terms in some asymptotic formulae in the theory of numbers—I".[50]

The mathematical significance of this paper has been nicely described by Prof. S. D. Adhikari in his article entitled "Some omega results and related questions in the research work of Pillai".[51]

It may be noted that Euler's totient function $\varphi(n)$ is an arithmetic function defined on natural numbers. It counts the number of natural numbers $1 \leq m \leq n$ with $(m, n) = 1$. It has been observed that the value of the said function fluctuates a lot. In analytic number theory, the study of the average behavior of this function is a classical problem. The methodology involving the determination of the main term, and the estimation of the error term is very vital. Because, if the error term is of the same order as the main term, nothing concrete can be deduced. It is noted that

$$\sum_{1 \leq n \leq x} \phi(n) = \frac{3}{\prod^2} x^2 + E(x),$$

---

[50] Section 1, (SSP. 9).

[51] "Collected works of S. S. Pillai". R. Balasubramanian and R. Thangadurai (Eds.). *Ramanujan Math. Soc.*, India.

where $E(x)$ is the remainder or error term in the average. When Pillai and Chowla started working on the Ω-estimate of the error term, the best available result was $E(x) = O(x \log x)$. Pillai and Chowla gave a far better estimate and found that $E(x) = \Omega$ $(x \log \log x)$. This result could be generalized after many years. Montgomery conjectured that $E(x) = O(x \log \log x)$ and $E(x) = \Omega(x \log \log x)$. But this is still an open problem, which is yet to be proved. Another result of Chowla and Pillai stated by them in the paper mentioned above is

$$\sum_{1 \le n \le x} E(n) \sim \frac{3}{2 \prod^2} x^2.$$

This is a result worth considering. Later, Chowla himself made further contributions in this area. In the fifties of the twentieth century, Erdös and Shapiro, Montgomery, and Pétermann carried out more research work in this direction.

In the next joint publication by Pillai and Chowla titled "On the error terms in some asymptotic formulae in the theory of numbers—II",[52] they followed up their studies of $E(x)$ for analogous questions in the case when totient function $\varphi(n)$ is replaced by $\sigma(n)$, the sum of the divisors of n. After further correspondence between these two great number theorists, Chowla and Pillai jointly published their third paper entitled "The number of representations of a number as a sum of n non-negative nth powers".[53] In this paper, they derived the omega result for $r_{n,n}(N)$ as $r_{n,n}(N) = \Omega$ $(\log \log N)$, where $r_{s,n}$ denotes the number of representations of N as a sum of s non-negative nth powers. The three results discussed above are involved with omega results.

Another type of problem that caught the attention of Pillai and Chowla had its roots in ancient Indian mathematics from the first millennium. From the time of Brahmagupta in the sixth century AD and Jayadeva around the eleventh century AD or maybe even earlier, it was known that continued fraction [12] is periodic. It was also known that all integral solutions of the equation $x^2 - Ny^2 = m$, $|m| < (N)^{1/2}$ were obtainable using the knowledge of continued fraction.

In 1927, T. Vijayaraghavan, a noted number theorist and a student of K. Ananda Rau, in his paper titled "Periodic simple continued fractions",[54] proved that the length L of the period of continued fraction satisfies the relation $L = O(N^{1/2} \log N)$. In 1931, Pillai and Chowla in their fourth joint paper titled "Periodic simple continued fractions"[55] proved that $L = O(N^{1/2} \log \log N)$ under generalized Riemann hypothesis [13]. In the correspondence between Pillai and Chowla, from 1935 onwards, one notices that gradually Chowla too was developing an interest in Waring's problem. So thereafter, Pillai and Chowla started to work jointly on topics related to Waring's

---

[52] Section 1, (SSP. 11).
[53] Section 1, (SSP. 30).
[54] Section 1, (TV. 1).
[55] Section 1, (SSP. 14).

problem. The outcome was the publication of their fifth and final joint paper titled "Hypothesis K of Hardy and Littlewood".[56]

In 20 long years, S. Chowla wrote more than 20 long letters to S. S. Pillai. They contained high-end mathematical discussions, which turned out to be very productive. The last letter of Chowla to Pillai was written in June 1950. A couple of months after that in August 1950, S. S. Pillai died in a plane crash near Cairo while on a proposed visit to the USA.

Getting back to Pillai's stay and researches at Annamalai University, it may be noted that during that period, he also worked on normal, highly composite numbers, arithmetic functions $\varphi(n)$, $d(n)$, and distribution of primes and primitive roots. After spending a decade in Annamalai University and a brief stint at Travancore University, in 1942, Pillai was made a lecturer at the prestigious Pure Mathematics Department of the Calcutta University. The famous German mathematician Prof. F. W. Levi who was the Head of the Department of Pure mathematics at that time had taken the initiative to get Dr. S. S. Pillai appointed at Calcutta University. During his brief sojourn there (1942–1945), Pillai published more than 15 research papers on highly composite numbers, arithmetic functions, primitive roots, etc.

In 1940, Pillai proved that any set of n consecutive positive integers, where $n \leq 16$, contains an integer which is relatively prime to all the others. However, there are infinitely many sets of 17 consecutive integers, where the above result of Pillai fails. Again, in his paper titled "On a linear Diophantine equation",[57] he showed that for any $m \geq 17$, there are infinitely many blocks of m consecutive integers where the above property does not hold. It may also be noted that in an earlier paper entitled "On the inequality $0 < a^x - b^y \leq n$",[58] Pillai had conjectured that given any number, one can find out consecutive terms whose difference is greater than the given number. However, this conjecture is still an open problem and has not yet been proved.

During his tenure at the University of Calcutta, S. S. Pillai developed an interest in the theory of functions of real variables. He was also trying to solve some unsolved problems in the theory of the Fourier series.

Another topic where Pillai made some notable contributions is on *smooth numbers*. Such numbers are defined as numbers which have only small prime factors. For example, $1620 = 2^2 \times 3^4 \times 5$. Since none of the prime factors are greater than 5, 1620 is called 5-*smooth*.

Pillai's personal contributions in number theory were more than 70 research papers. But during his brief stay at Calcutta University, he trained up a brilliant student named L. G. Sathe. In a way, this was a phenomenal contribution to the

---

[56] Section 1, (SSP. 30).
[57] Section 1, (SSP. 54).
[58] Section 1, (SSP. 14).

Bengal school of research on number theory. This episode will be taken up during the detailed discussions of the Bengal school.

Annexure

[10] *Exponential Diophantine Equations*: Any equation in one or more variables having solutions in integers is called a *Diophantine equation*. It is named after the famous Greek mathematician Diophantus, as he first initiated the study of such equations. An exponential Diophantine equation is one in which exponents on terms can be unknowns, or one may say an equation in which bases are (given or unknown) integers. The exponents are unknown *integers*. Such equations have an additional variable or variables occurring as *exponents*. For example, equations of the type

$$1 + 2^a = 3^{b+}5^c, 3^a + 7^c = 3^c + 5^d + 2.$$

The Ramanujan–Nagell equation of the type $2^n - 7 = x^2$ is an exponential Diophantine equation.

[11] *Omega results*: An Ω-*result* (omega result) is an expression of the form $f(x) = \Omega g(x)$, where $g(x) > 0$ for sufficiently large x. It means that there is an $A > 0$ such that $|f(x)| > Ag(x)$ for an unbounded sequence of real numbers.

[12] *Continued fraction*: It is a fraction of infinite length whose denominator is a quantity plus a fraction, whose latter fraction has a similar denominator, and so on.

[13] *Riemann hypothesis*: The Riemann hypothesis is one of the most important conjectures in mathematics. It is a statement about the zeros of Riemann zeta function. When the Riemann hypothesis is formulated by Dirichlet's L-functions, then it is known as *generalized Riemann hypothesis* (GRH).

2.1.4 T. Vijayaraghavan (1902–1955)

A product of the University of Madras and a student mentored by K. Ananda Rau, Tirukkannapuram Vijayaraghavan was born in a village in the South Arcot district of erstwhile Madras Presidency (now Tamil Nadu). His father was a noted scholar of Tamil and Sanskrit. Vijayaraghavan probably inherited his passion for intellectual pursuits from his father. However, right from his school days, mathematics was his greatest passion.

After completing his school education, he completed his intermediate examination in 1920. Thereafter, he studied in the famous Presidency College of Madras from 1920 to 1924 but was unable to obtain the honors degree in mathematics. The explanation for this failure is stated as follows by K. Chandrasekharan:

This is not altogether surprising, since he (Vijayaraghavan) was genuinely interested, even at that stage, in serious mathematics, and had already begun thinking about research problems which could have stumped many of his teachers.

However, the great mentor of students, Prof. K. Ananda Rau had recognized Vija-yaraghavan's talent quite early and was instrumental in getting him admitted to Presidency College in Madras. In later years, Vijayaraghavan remembered his mentor and teacher Prof. K. Ananda Rau with nostalgia and affection. He admitted that Prof. Rau had inspired him and had drawn him to serious mathematics. In 1920, even before completing his intermediate course, Vijayaraghavan had successfully published a research note titled "On the set of points $\{\varphi(n)/n\}$".[59] The paper dealt with the limit points of the set $\{\varphi(n)/n\}$, where Euler's function $\varphi(n)$ is the number of positive integers less than n and prime to n.

Around 1921, clearly influenced by the stories and achievements of Srinivasa Ramanujan, Vijayaraghavan started sending some of his results and manuscripts to Prof. G. H. Hardy in England. Initially, Hardy did not respond. Then one of Vijayaraghavan's former teachers Prof. S. R. Ranganathan from Madras Presidency College, while on a visit to England, met Prof. Hardy at Oxford; Hardy talked to him and Prof. Ranganathan showed him a research paper by Vijayaraghavan. After that, Hardy took a very swift action in a very unconventional way. The result was Vijayaraghavan's entry into the New College at Oxford. The episode has been interestingly presented by Dr. S. R. Ranganathan. According to him:

> ...For, in March, 1925, I personally saw him (Hardy) making a similar decision with spontaneity and certitude. He was the Sevillian Professor of Geometry in the New College, Oxford. I met him at that time. I also had lunch with him. My main purpose was to place before him a paper on Tauberian Theorem by T. Vijayaraghavan who was a gifted student of the Department of Mathematics in the Presidency College, when I was teaching there. After perusing this paper, Hardy stood up with a jerk and said, 'This is a remarkable mathematician from your country—next only to Ramanujan. He should be brought over to me'.

> This he said even without waiting to know any details about Vijayaraghavan. I told him that he had failed in the Honours Examination just the previous year. Hardy asked how it happened. I said that Edward B. Ross—my own professor and one of his old students—and myself were examiners and that in spite of our personal knowledge of his extraordinary mathematical ability we could not even give him pass marks, as he had not answered most of the questions in all but one paper, viz., the one on Functions of a Complex Variable. After hearing this, Hardy remarked, 'It makes no difference'. ...Hardy decided in a trice to get T. Vijayaraghavan to Oxford.

So in 1925, with a scholarship from the University of Madras, T. Vijayaraghavan finally joined the New College at Oxford and started working under the guidance of Prof. G. H. Hardy. He spent three fruitful years, during 1925–1928 at Oxford and published several good research papers.

In number theory, he first published the research paper titled "Periodic simple continued fractions".[60] It may be mentioned in this context that Vijayaraghavan had developed an interest in continued fractions from his Honors class days in the Madras Presidency College. Famous French mathematician Lagrange, while proving classical results that every quadratic surd corresponds to a periodic, simple, continued fraction, had shown that if a quadratic surd is $(P + R^{1/2})/Q$, where $|P|$, $|Q|$, R are

---

[59] *Journ.Indian Math. Soc.*, 12, (1920), 98–99.

[60] Section 1, (TV. 1).

the least positive integers, then the number of elements in the periodic part of the corresponding simple continued fraction is less than $2R$. Vijayaraghavan improved this estimate. According to the famous number theorist K. Chandrasekharan:

Vijayaraghavan improved this estimate by showing in an elementary way that the $2R$ can be replaced by $O(R^{1/2} + \varepsilon)$, $\varepsilon > 0$.... Associated with this result is another result by him on the solvability of the Pellian equation $Rx^2 = y^2 + 1$. It is well known that this equation is, or is not, solvable in integers according as the continued fraction for $R^{1/2}$ has, or has not, a period with an odd number of elements. Vijayaraghavan considered the more general surd $(P + R^{1/2})/Q$, and showed that it has, or has not, a period with an odd number of elements, according as the Pellian equation is, or is not, solvable for R.

Vijayaraghavan's next paper was titled "A note on Diophantine approximation".[61] Hardy and Littlewood had written a famous memoir on Diophantine approximation [14].[62] Vijayaraghavan's paper mentioned above was based on their results. Hardy and Littlewood in their memoir had shown that if (x) denote the fractional part of x, and $\varphi = \varphi(\lambda, \theta, \alpha)$ denote the least positive integer n, the approximation $|(n\theta) - \alpha| < 1/\lambda$ holds. Using a conjecture of Hardy and Littlewood, Vijayaraghavan showed that for specific choices of θ and α, φ could be made arbitrarily large, for a sequence of values of λ tending to infinity. Kronecker's fundamental theorem on Diophantine approximation suggests the plausibility of a similar result with any number of θ's. Taking the advantage of this concept, Vijayaraghavan considered the case of two θ's and proved somewhat different but interesting results.

After returning to India from England, in 1928, Vijayaraghavan took up an appointment at Annamalai University for one year. In 1930, he was selected and joined as a lecturer and joined Aligarh Muslim University (AMU). There he came in contact with the world famous French mathematician André Weil. André Weil was invited to work at AMU and he in turn selected T. Vijayaraghavan as a lecturer to work with him at the Department of Mathematics. Dr. Vijayaraghavan was very close to Prof. Weil and both were brilliant mathematicians. In 1931, the AMU authorities in a very autocratic and arbitrary manner terminated the services of Prof. Weil and offered the post to Dr. Vijayaraghavan. Vijayaraghavan, however, took great offense in the happenings. As a mark of protest against the shabby treatment and injustice meted out to Prof. Weil, he too resigned from AMU. In the same year, in 1931, he joined the Mathematics department of Dacca University in erstwhile East Bengal (today, Bangladesh) in undivided India as a reader. He worked there until 1946. In 1946, he resigned and joined as a professor at Andhra University, Waltair. In 1949, he left Andhra University to become the director of the privately owned Ramanujan Institute of Mathematics in Madras.

During his tenure at Dacca, T. Vijayaraghavan was responsible for initiating research on number theory in that obscure university of undivided Bengal. Detailed discussions on this will be done later.

---

[61] Section 1, (TV. 2).

[62] *Acta Mathematica*, 37, (1914), 155–190.

While teaching at the Dacca University, Vijayaraghavan worked on mathematical analysis as well as number theory. In 1936, Vijayaraghavan at the initiative of American mathematician G. D. Birkhoff was invited to the USA as a visiting lecturer of the American Mathematical Society. After returning from the USA, he continued his work on mathematical analysis.

However, in 1939 and 1940, he worked seriously on problems related to number theory and published the following two papers:

- "On the irrationality of a certain decimal" [Sect. 2.1, (TV. 3)]
- "On decimals of irrational numbers" [Sect. 2.1, (TV. 4)].

Thereafter, Dr. T. Vijayaraghavan became deeply involved in the study of fractional powers of real numbers. According to the noted number theorist H. Davenport, Vijayaraghavan's important contribution was as follows:

> ...concerned the distribution of fraction part of θ^n, where $\theta > 1$ and $n = 1, 2, 3,....$ G. H. Hardy had earlier proved that if θ is an algebraic number, and if there exists $\lambda > 0$ such that the fractional part of $\lambda \theta^n$ has only the limit points 0 and 1 as n tends to infinity, then θ is an algebraic integer and all the algebraic conjugates θ' of θ (other than θ itself) satisfy $|\theta|<1$.

Vijayaraghavan published a series of four papers on the fractional parts of powers of real numbers, the first one being

"On the fractional parts of the powers of a number, I" [Sect. 2.1, (TV. 5)].

The second paper titled

"On the fractional parts of the powers of a number, II" [Sect. 2.1, (TV. 7)]

is considered to be most substantial. In this paper, he proved a generalization of Hardy's result. He showed that if $\theta_1, \theta_2, ..., \theta_r$ are distinct algebraic numbers with $|\theta_i| < 1$, and if there exist $\lambda_1, \lambda_2, ..., \lambda_r$ (none of them zero) such that the sequence formed by the fractional parts of $\lambda_1 \theta_1{}^n + \lambda_2 \theta_2{}^n + \cdots + \lambda_r \theta_r{}^n$ ($n = 1, 2, 3, ...$) has only a finite number of distinct limit points, then each of $\theta_1, \theta_2, ..., \theta_r$ is an algebraic integer with all its algebraic conjugates of absolute value less than 1. The algebraic integers greater than 1 with conjugates absolutely less than 1 are called *Pisot–Vijayaraghavan numbers* (since M. Pisot had independently encountered them).

The international impact of Pisot–Vijayaraghavan numbers is considerable. Among others, Salem and Siegel have discovered more about these numbers. Some problems in the area still remain unsolved. For example, it is not known, whether, if $\theta > 1$ is any real number not in the class, the fractional part of θ^n is everywhere dense in $(0, 1)$.

The remaining two papers in this series by Dr. T. Vijayaraghavan are

- "On the fractional parts of the power of a number, III" [Sect. 2.1, (TV. 8)]
- "On the fractional parts of the power of a number, IV" [Sect. 2.1, (TV. 12)].

Apart from these publications, his number theory related to other papers are based on different topics. He published a paper on the general rational solution of some particular types of Diophantine equations. It was published in the *Proceedings of the Academy of Sciences*.[63] He published another paper on Jaina magic squares.[64]

Dr. Vijayaraghavan was a close friend of both S. S. Pillai and S. Chowla. He had regular correspondence with them. The contents of those letters were mainly mathematical. In collaboration with S. Chowla, T. Vijayaraghavan wrote the following three papers. They are

- (With Chowla, S.) "The complex factorization (mod p) of the cyclotomic polynomial [15] of order $p^2 - 1$" [Sect. 2.1, (TV. 9)]
- (With Chowla, S.) "Short proofs of theorems of Bose and Singer" [Sect. 2.1, (TV. 10)]
- (With Chowla, S.) "On Complete residue sets" [Sect. 2.1, (TV.13)].

His contributions as a top number theorist of India and his initiation of a school of research on number theory in Dacca University of the undivided Bengal will always be remembered with admiration. Dr. T. Vijayaraghavan passed away in Madras of a massive heart attack in 1953.

Annexure

[14] *Diophantine approximation*: In number theory, the area of Diophantine approximation deals with the approximation of a real number by natural numbers.

[15] *Cyclotomic polynomial*: Cyclotomic polynomials are polynomials whose complex roots are primitive roots of unity.

2.2   The Panjab School of Research on Number Theory (1920–1999)

One can safely and unhesitatingly say that the founding father of the Panjab School of research on number theory was Sarvadaman Chowla. He was a close associate of both S. S. Pillai and T. Vijayaraghavan. A brief discussion on his role as a number theorist, a mentor of the young scholars, and the international impact of his research is discussed now.

2.2.1   Sarvadaman Chowla (1907–1995)

Sarvadaman Chowla (or S. Chowla) was born in 1907 in London. S. Chowla's father Gopal S. Chowla (G. S. Chowla) was a famous mathematician and went to Trinity

---

[63] Sect. A, 12, (1940), 284–289.
[64] Section 1, (TV. 6).

College, Cambridge, U.K., to pursue higher studies. During that time, S. Chowla was born. On his return to India, G. S. Chowla was made the senior professor of Mathematics at the Government College in Lahore and subsequently he was promoted to the prestigious Indian Educational Service. Young S. Chowla was then pursuing his studies in Lahore and showing great aptitude for mathematics. In fact, at the age of 18, he published his first research paper titled "Some results involving prime numbers".[65] That was the beginning of his journey as a research worker in number theory. In 1928, he obtained his master's degree from the Government College in Lahore. A year later, young Chowla's father G. S. Chowla decided to admit his son into Trinity College of Cambridge University. So, in 1929, Prof. G. S. Chowla took a long leave and took his son to England and admitted him to Trinity College. After that, he with the other members of his family went on a holiday to Paris. Unfortunately, Prof. G. S. Chowla was infected with pneumonia and in December 1929 he passed away in Paris. In spite of this great tragedy at the very beginning of his research career, S. Chowla started working under the guidance of Prof. J. E. Littlewood from 1929. S. Chowla was given the freedom to follow his own interests and in their infrequent meetings, Littlewood encouraged him to carry on research in his own way.S. Chowla worked on "k-analogue of Riemann zeta function" and "hypothesis k" of Hardy and Littlewood (for Waring's problem). In 1931, he completed his thesis titled "Contributions to the analytic theory of numbers" and obtained the Ph.D. degree from the University of Cambridge. It may be interesting to note that young Chowla had submitted a list of 30 published papers along with his thesis. After he obtained his Ph.D. degree, three chapters from his dissertation were published. They were.

- "Two problems in the theory of lattice points" [Sect. 2.2, (SC. 28)]
- "Contributions to the analytic theory of numbers" [Sect. 2.2, (SC. 29)]
- "Contributions to the analytic theory of numbers" [Sect. 2.2, (SC. 31)].

In 1934, S. Chowla was awarded the "Ramanujan Memorial Prize", initiated by the University of Madras for the best original thesis in mathematics. In 1931, S. Chowla after completing his studies in Cambridge returned to India and joined the Mathematics Department of St. Stephen's College in Delhi. In 1932, he shifted to the Benares Hindu University (BHU) at Benares and joined as a lecturer there. In 1933, at the persuasion of Dr. S. Radhakrishnan, the then vice-chancellor of Andhra University at Waltair, he joined the university as a reader and took charge as the Head of the Department of Mathematics. After spending three years there, he finally, in 1936, joined the Government College of Lahore as a professor of mathematics. For him, it must have been a nostalgic and sentimental event as it was both his *alma mater* as well as the department where his late father was an eminent faculty member.

    S. Chowla's research contributions in the area of number theory are quite remarkable. Especially his 20-year long correspondence with another great number theorist S. S. Pillai, as well as his stay at Benares, Waltair, and Lahore had a great impact

---

[65] Section 2, (SC, 1).

on Indian mathematics and Indian mathematicians. Apart from S. S. Pillai, T. Vija-yaraghavan also collaborated with S. Chowla and jointly solved important problems on number theory. Chowla was a great teacher and an inspiring research guide. One of his doctoral students referred to him as a "perpetual ambassador for number theory". Chowla was the man who trained up number theorists like A. Sreerama Sastri of Andhra University, Waltair, F. C. Auluck, Mian Abdul Majid, Daljit Singh, and R. P. Bambah of Panjab University. He also collaborated with Hansraj Gupta of Panjab University. Undoubtedly, S. Chowla played a leading role in building up a strong school of research on number theory in Panjab University, which later came to be known as Panjab School of Research on Number Theory.

In order to assess Chowla's contributions to the development of the school of research on number theory, it is necessary to take into account his own research contributions as well as his collaborative work with other number theorists and his famous students.

Among his Indian collaborators with whom S. Chowla wrote joint papers, the names include K. Ananda Rau, S. S. Pillai, S. Sastry, F. C. Auluck, Mian A Majid, R. C. Bose, C. R. Rao, D. Singh, A. R. Nazir, T. Vijayaraghavan, R. P. Bambah, and D. B. Lahiri. S. Chowla was the man who inspired and encouraged F. C. Auluck and R. P. Bambah to take up the career of doing research in mathematics. S. Chowla's brother (late) Inder Chowla and daughter Paromita Chowla were drawn to mathematical research also probably inspired by his unwavering devotion to the subject.

Between the years 1925–1931, S. Chowla solved 34 problems which were posed in the *Journal of the Indian Mathematical Society*. He published the following five papers in the *Journal of the Indian Mathematical Society* during 1925-1926, which were all completed even before he finished graduation:

- "A new proof of Von Staudt's theorem" [Sect. 2.2, (SC. 2)]
- "Solution of Q. 1084, 1086 (Hemraj)" [Sect, 2, (SC. S7)
- "Some results involving prime numbers" [Sect. 2.2, (SC. 1)]
- "Remarks on Q. 353 of S. Ramanujan" [Sect. 2.2, (SC. S5)]
- "Solution of Q. 1070 of S. Ramanujan" [Sect. 2.2, (SC, S6)].

In the early years of his academic career, S. Chowla, like many other mathematicians of India, was greatly influenced by Srinivasa Ramanujan and his fabulous legacy. Even while solving the problems posed in the *Journal of the Indian Mathematical Society*, Chowla had a special fascination for solving problems which were earlier posed by S. Ramanujan. Here, a few such examples are cited. In the *Journal of the Indian Mathematical Society*,[66] he solved a problem set by S. Ramanujan. Again in the same volume of the journal, Chowla in collaboration with N. B. Mitra and S. V. Venkataraya Sastri solved another problem,[67] which was also posed by S. Ramanujan. Both these problems were related to elementary number theory. In 1928, Chowla in collaboration with T. R. Raghavasastri, T. Totadari Aiyangar, and T. Vijayaraghavan

---

[66] Section 2, (SC. S7).
[67] Section 2, (SC. S6).

solved a problem set by K. Ananda Rau.[68] This was Chowla's first interaction with Vijayaraghavan. The problem was on prime numbers. In the same issue of the journal, he solved two more problem set by S. Ramanujan. The first one was related to elliptic functions in number theory.[69] The second one was also related to number theory.[70] In 1929, Chowla solved a problem on prime numbers posed by K. Ananda Rau.[71] In 1930, Chowla solved two problem set by S. S. Pillai. Both were related to number theory.[72] This was probably Chowla's first academic interaction with S. S. Pillai. In 1931, S. Chowla in collaboration with Budharam, Hukam Chand, K. K. Vedantham, N. P. Subramaniam, and Hansraj Gupta solved yet another problem on number theory.[73] Apart from these solutions to problems posed by mathematicians of renown, S. Chowla independently published two research papers in 1926. The first one dealt with prime numbers. It was titled "Some results involving prime numbers".[74] The second paper related to Bernoulli numbers was titled "A new proof of Van Staudt's theorem."[75] In 1927, S. Chowla published a series of papers related to number theory. The notable ones are

- "An elementary treatment of the modular equation of the third order" [Sect. 2.2, (SC. 4)]
- "On the order of d (n), the number of divisors of n" [Sect. 2.2, (SC. 5)].

Chowla's research paper on Eulerian and Bernoullian numbers titled "Some properties of Eulerian and Bernoullian numbers", was published in *Messenger of mathematics*, England [Sect. 2.2, (SC. 6)]. This was his first publication in an international journal.

The 20-year long correspondence between S. Chowla and S. S. Pillai and the resulting output has been discussed earlier in detail in the context of S. S. Pillai. So, that is not repeated here again.

Attention should now be focused on S. Chowla's influence and efforts to guide talented young mathematicians and enrich research activities on number theory in India. According to celebrated number theorist Prof. R. Balasubramanian:

> Perhaps spurred by the exciting results of his friend and collaborator S. S. Pillai, Sarvadaman Chowla encouraged Faqir Chand Auluck (F. C. Auluck) to apply the method of Vinogradov as modified by Gelbcke to Waring's problem for solving the problem.

A brief introduction of F. C. Auluck (1912–1987) would be relevant here. F. C. Auluck was born in Jalandhar, Panjab, in 1912. He had a uniformly brilliant academic career

---

[68] Section 2, (SC. S13).

[69] Section 2, (SC. S14).

[70] Section 2, (SC. S16).

[71] Section 2, (SC. S22).

[72] Section 2, (SC. S27 and S28).

[73] Section 2, (SC. S33).

[74] Section 2, (SC. 1).

[75] Section 2, (SC. 2).

and had stood first in first class in both B.A. and M.A. examinations from the Panjab University in Lahore. In the thirties of the twentieth century, he took up the job of a lecturer at the Dayal Singh College of Lahore in undivided Panjab. He simultaneously started doing research on number theory under the guidance of Prof. S. Chowla. In 1937, he in collaboration with Chowla published a paper entitled "A property of numbers".[76] Auluck also wrote a brief paper on geometry related to Simson line, in the same issue of the journal mentioned above. Later in the same year, S. Chowla published a paper titled "Auluck's generalization of the Simson line property".[77] Another paper jointly written by Chowla and Auluck was titled "The representation of a large number as a sum of 'almost equal' squares".[78] In 1940–1941, F. C. Auluck published three research papers in collaboration with S. Chowla:

- "An approximation connected with exp x" [Sect. 2.2, (SC. 96)]
- "On Weierstrass' approximate theorem" [Sect. 2.2, (SC. 97)]
- "Some properties of a function considered by Ramanujan" [Sect. 2.2, (SC. 98)].

Again in collaboration with S. Chowla and Hansraj Gupta, Auluck published a paper titled "On the maximum value of the number of partitions of n into k parts".[79]

But perhaps Auluck's most notable contribution that year was on Waring's problem. Around that point of time, S. Chowla perhaps influenced by his friend and long-time collaborator S. S. Pillai had started taking an active interest in Waring's problem. So, he encouraged not only Auluck but another research scholar named Inder Chowla to look into Waring's problem. He also published some papers on Waring's problem and congruences. F. C. Auluck showed that every natural number n such that $\log_{10}\log_{10} n \geq 89$ is indeed a sum of biquadrates. This work was independently published by Auluck as "On Waring's problem for biquadrates".[80]

In the introduction to the paper, Auluck himself admitted that his bound was extremely difficult to compute, so that improvements on the upper bound for $g(4)$ was practically impossible. For a long time, Auluck's upper bound was the only explicit bound known on the representation for 19 biquadrates.

F. C. Auluck continued to work on his own, and also in collaboration with S. Chowla, on problems in number theory. But subsequently, he switched over to physics. He became a famous physicist famed for his work in statistical physics. He moved over to New Delhi from Lahore. Eventually, he became Aryabhata Professor and later Emeritus Professor of physics at the University of Delhi. F. C. Auluck passed away in 1987 at the age of 75.

The other scholar of S. Chowla was his brother Inder Chowla who has already been mentioned before. He published a paper titled "On Waring's problem in cubes".[81]

---

[76]Sect, 2, (SC. 89).

[77]Section 2, (SC. 92).

[78]Section 2, (SC. 93).

[79]Section 2, (SC. 99).

[80]*Proceed. Indian Acad. Sci.*, Sect. A, 11, (1940), 437–450.

[81]*Proc. Indian Acad. Sci.*, Sect. A, (1937), 1–17.

In this paper, he proved that almost all numbers are expressible as a sum of four non-negative cubes (of integers), provided a certain hypothesis was satisfied. In the same issue of the journal, I. Chowla published another research paper titled "On the number of solutions of some congruences in two variables".[82] This was an interesting paper. The author mentioned in the beginning that Hurwitz gave the first proof that the congruence $ax^k + b^k \equiv O(p)$ is solvable with x, y prime to p, whenever a, b, c $\neq O(p)$, for sufficiently large p. This was a generalization of an earlier result due to Dickson. Mordell, Davenport, and Hasse proving the stronger result that the number of solutions of the above congruence is asymptotic to p, for large p. On the other hand, I. Chowla in his paper gave a simple proof of this result by a new method based on the lemmas due to Hardy and Littlewood.

S. Chowla's other collaborators were Hansraj Gupta and R. P. Bambah who became pioneer number theorists of the Panjab School. But this will be discussed later. Getting back to S. Chowla, the partition of India prior to Independence in 1947 changed the course of life for him. After August 15, 1947, S. Chowla shifted to Delhi for a while. But in 1948, he left India and took up the job of a temporary lecturer at the Institute of Advanced Study in Princeton, U.S.A. There once again Chowla became actively involved in research on topics related to number theory. His colleagues there were world famous number theorists like Carl L. Siegel, Atle Selberg, Paul Turan, Paul Erdös, and Hermann Weyl. In 1949, he joined the University of Kansas in Lawrence, U.S.A. Three years later, he moved to the University of Colorado in Boulder, U.S.A. Finally, in 1963, he joined as a research professor at Pennsylvania State University and served there until his retirement in 1976. During his years in the U.S.A., he was actively involved in research on various topics of number theory. During his long research career, which started in 1925 and ended in 1986, he wrote a total of 350 research papers. During his long years of service, in the U.S.A., Chowla guided more than 20 Ph.D. scholars who obtained their degrees under his supervision. The Indian students he guided there include his own daughter P. Chowla, Sahib Singh, and A. M. Vaidya. While working in different universities, S. Chowla collaborated with Atle Selberg, Paul Erdös, L. J. Mordell, H. Davenport, E. Artin, R. Brauer, N. C. Ankeny, P. T. Bateman, G. Shimura, H. J. Ryser, I. N. Herstein, D. Lewis, H. Hasse, B. W. Jones, T. M. Apostol, E. Strauss, H. B. Mann, T. Skolem, B. J. Birch, Marshall Hall Jr., A. Borel, K. Iwasawa, R. Ayoub, H. Zassenhaus, B. C. Berndt, K. Ramachandra, and many others.

It would not be an overstatement to say that he worked in collaboration with the brightest number theorists of his time. At a conservative estimate, he wrote more than 300 papers with about 60 co-authors. He instilled the love of mathematics in general and number theory in particular into many lives.

In additive number theory [16], he worked on lattice points [17], partitions, and Waring's problem. In analytic number theory, S. Chowla had publications on Dirichlet L-functions [18], primes, and Riemann and Epstein zeta functions [19]. He also made notable contributions on binary quadratic forms [20] and class numbers [21], Diophantine equations, and Diophantine approximations.

---

[82]*Proceed. Indian Acad. Sci.*, Sect. A, 40–44.

Again getting back to the Indian scenario, it may be noted that in 1944, he in collaboration with R. C. Bose and C. R. Rao published a paper titled "A chain of congruences".[83] Subsequently in the same year, in collaboration with R. C. Bose, another paper titled "On a method of constructing a cyclic subgroup of order $p + 1$ of the group of linear fractional transformations (mod p)"[84] was published. During this period, S. Chowla also collaborated with Mian Abdul Majid and they together published three papers. But their interest was in mathematical analysis and differential equations. Details of those publications are not relevant here. Later, after the partition of the Indian subcontinent, Mian Abdul Majid stayed back in Lahore. Toward the end of the year, the trio of S. Chowla, R. C. Bose, and C. R. Rao published a paper titled "On the integral order (mod p) of quadratics $x^2 + ax + b$, with applications to the construction of minimum functions for $GF(p^2)$, and to some number theory results".[85] In this paper, they considered polynomials whose coefficients belonged to the ring of integers and p always denoted an odd prime. Again if n is taken to be the least positive integer such that $x^n \equiv$ an integer (mod p, $x^2 + ax + b$), then n is called the *integral order* of $x^2 + ax + b$, (mod p). In this paper, they deduced a number of interesting theorems about the "integral order". In the same year, S. Chowla and R. C. Bose jointly published the paper entitled "On the construction of affine difference sets".[86] The three mathematicians S. Chowla, R. C. Bose, and C. R. Rao again collaborated and they published two papers:

- "A chain of congruences" [Sect. 2.2, (SC. 122)]
- "Minimum functions in Galois fields" [Sect. 2.2, (SC. 118)].

This was the last published paper by this famous trio.

The discussion on the interactions of S. Chowla and his famous student R. P. Bambah would be carried out in detail while focusing on the research contributions of R. P. Bambah. At present, a brief summary of Chowla's research work is presented below, indicating the topics involved.

Chowla did a substantial amount of research on Ramanujan's $\tau(n)$ functions. During 1946–1947, Chowla and his famous student R. P. Bambah proved congruences for $\tau(n)$ modulo various powers of 2, 3, 5, 7. In 1947, they showed that

$$\tau(n) \equiv \tau\left(\bmod\ 2^5 \cdot 3^2 \cdot 5^{2.7} \cdot 23 \cdot 691\right) \text{for almost all } n.$$

Chowla extended it and deduced a notable result. He stated that:

Let a, b, c, d, e, f be arbitrary integers ≥ 0. Then $\tau(n) \equiv 0 \pmod{2^a \cdot 3^b \cdot 5^c \cdot 7^d \cdot 23^e \cdot 691^f}$ for almost all n.

---

[83] Section 2, (SC. 122).

[84] Section 2, (SC. 123).

[85] Section 1, (SC. 116).

[86] *Bulletin of the Calcutta Mathematical Society*, 37, (1945), 107–112.

S. Chowla's next important work is related to Margulis–Oppenheim theorem. Alexander Oppenheim (1903–1994) was a British number theorist and a student of G. H. Hardy and L. E. Dickson (1874–1954). Influenced by them, he worked on Diophantine approximation and in 1929, he made a conjecture. His conjecture was concerned with the representation of numbers by real quadratic forms in several variables. In 1934, Chowla made the first breakthrough in the said conjecture. In 1946, H. Davenport (1907–1969) and H. A. Heilbronn (1908–1975) extended Chowla's result. Through the efforts of Davenport, Birch, and Ridout, in the 1950s, Oppenheim's conjecture was proved for general indefinite quadratic forms in $n \geq 21$ variables. After more than 30 years, in 1987, G. Margulis completely proved the Oppenheim conjecture by using ideas from topological groups. It may be noted that Margulis was influenced in his proof by the ideas of two famous Indian mathematicians namely M. S. Raghunathan and S. G. Dani. Apart from Raghunathan and Dani, K. G. Ramanathan, S. Raghavan, and Gopal Prasad have also made significant contributions to related problems.

Another area where Chowla's work has been much admired concerns his work on class numbers. In fact, he seemed to have a lifelong fascination for class numbers. During his tenure in the U.S.A., Chowla in collaboration with N. C. Ankeny (1927–1993) and E. Artin (1898–1962) proved many notable results for the class numbers of quadratic fields and deduced interesting results.

Chowla used Hansraj Gupta's partition function related tables to disprove one of Ramanujan's conjectures in partition theory. The details will be reported while dealing with Hansraj Gupta's contributions in number theory.

Chowla's contributions during his stay in India and with Indian mathematicians resident in India have been more or less brought to the fore. But the prolific researcher that he was, he went on doing significant research and publishing important results throughout his productive years in the U.S.A. His name is identified with a number of mathematical results with foreign mathematicians as his collaborators. For the sake of completeness, a few more important such results are mentioned. The Bruck–Chowla–Ryser theorem gives a criterion for the non-existence of certain block designs. The Ankeny–Artin–Chowla congruence for the class number of a real quadratic field has already been mentioned before. The Chowla–Mordell theorem on Gauss sums is another important contribution. In 1967, Selberg and Chowla discovered a beautiful formula, well-known as the Chowla–Selberg formula which is related to special values of the Dedekind eta function.

Appendix

[16] *Additive number theory*: This special type of number theory which is called *additive number theory* studies subsets of integers and their behavior under addition.

[17] *Lattice points*: A point *lattice* is a regularly spaced array of points. Formally, a lattice is a discrete subgroup of Euclidean space, assuming it contains the origin.

[18] *Dirichlet L-function*: If X is a Dirichlet character and is a complex variable and its real part is >1, then by analytic continuation the function can be extended

to a meromorphic function on the whole complex plane. It is then called a
Dirichlet L-function and is denoted by $L(S, X)$.
Riemann zeta function: Riemann zeta function is

$$\zeta(s) = 1 + \frac{1}{2^s} + \frac{1}{3^s} + \cdots = \sum \frac{1}{n^s}$$

where the summation over n is from 1 to infinity and $\zeta(s)$ is a function of
a complex variable s. It is an important special function of mathematics and
physics that arises in definite integration and is intimately connected with deep
results surrounding the prime number theorem.
Epstien zeta function: It is a function belonging to a class of Dirichlet series
generalizing the Riemann zeta function. It is related to number theoretical
problems.

[19]   *Riemann and Epstein zeta functions*:
[20]   *Binary quadratic form*: A *binary quadratic form* is a quadratic form in two
       variables having the form $Q(x, y) = ax^2 + 2bxy + cy^2$. It is mathematically
       denoted by $<a, b, c>$.
[21]   *Class numbers*: The *class number* of an order of a quadratic field with
       discriminant is equal to the number of reduced binary quadratic forms of the
       discriminant.

The next important number theorist who made remarkable contributions and is
considered as an icon of the Panjab School is Hansraj Gupta. He was not a student
of S. Chowla, nor was he a collaborator, to start with. But Chowla used one of
Hansraj Gupta's results from the latter's tables on partitions and disproved one of
Ramanujan's conjecture. This gave international recognition to Hansraj Gupta. It
is time to introduce Hansraj Gupta and discuss his pioneering contributions in the
development of the school of number theory at Panjab University.

2.2.2   Hansraj Gupta (1902–1988)

Though both Profs. Hansraj Gupta (Fig. 2.4) and R. P. Bambah were born before
India attained her freedom from the colonial rulers, their main research activities
and particularly their contributions in building up the post-Independence School of
Number Theory in the newly setup Panjab University in Chandigarh are remarkable
from the historical point of view. So, roughly they have been considered in the second
half of the twentieth century.

Hansraj Gupta was born on October 9, 1902, at Rawalpindi (now in Pakistan). His
father Jati Ram Gupta was a petty official in erstwhile Patiala State. Hansraj Gupta
spent practically his entire childhood and did his school education there. In 1919,
he passed the matriculation examination from Panjab University, Lahore, obtaining
high first division marks and also winning the university scholarship. After that,

he joined the Mohindra College at Patiala to pursue further education. In 1921, he passed the intermediate examination again with high first division marks and won the Patiala State Scholarship. During the next two years, he studied mathematics for the undergraduate course in the same college. Profs. Bhagat Ram and Durga Dass Kapila taught him mathematics and, in 1923, he graduated with high second division. Teaching for a postgraduate course in mathematics was available only in Lahore. For most of 1923–1924, Hansraj Gupta was not well. Financial difficulties were also there. So, he could not enroll in a college at Lahore immediately. Fortunately, for the ailing Hansraj, there was some good news. Mohindra College at Patiala was affiliated to Panjab University for M.A. in mathematics but it did not have any arrangements for teaching. So, Hansraj enrolled there for the M.A. course and he did not have to attend the college as no teaching was available. In the process, he saved a year. In 1924, he joined Dyal Singh College at Lahore for the second year of the M.A. course. The Principal of the college was Pandit Hemraj who had a high reputation both as a mathematician and as a teacher. Hemraj had also published some papers on number theory. Probably, those were the basic reasons for Hansraj to choose that particular college. Pandit Hemraj was a well-known mathematician of those days. He had been associated with the Indian Mathematical Society (IMS) and was a member of the executive committee of the Society. In 1923, Hansraj Gupta was elected a member of IMS. While seeking migration to Dyal Singh College, Hansraj had mentioned about his membership of the IMS. That impressed Prof. Hemraj, and he got easily admitted there. In those days, the postgraduate teaching of mathematics was jointly done by three colleges at Lahore. The university professors also took an active part in the teaching process. At that time, Prof. G. S. Chowla (father of S. Chowla) was one of his teachers. In 1925, Hansraj Gupta passed the M. A. examination in mathematics with a high second division and stood first in the university. Considering the fact that he had received formal instructions only for a year, it was indeed a very creditable performance.

Before discussions on his formal academic career are recorded, it may be relevant to mention that Hansraj Gupta had a special aptitude for mathematics right from his school days. While still in school, he had devised his *100-year calendar*. As an undergraduate student, he had devised his *perpetual calendar*. In 1923, it was exhibited at the *British Empire Exhibition*, held at Wembley in suburban London. Hansraj Gupta was awarded a medal and a certificate of merit for the same.

Anyway, as jobs were difficult to get in those days in India under the rule of a colonial power, after completing his M.A. Hansraj Gupta practically sat idle for almost a year. For a few months, he tutored the son of the Prime Minister of Patiala. Finally, in August 1926, he joined the Sadiq Egerton College at Bhawalpur (now in the Sind province of Pakistan). After working there for two years, he got an offer from the Principal, Government Intermediate College at Hoshiarpur. He readily accepted the offer. In 1928, he joined there and his designation was Teacher of mathematics. In 1947, the college became a full-fledged degree college and Hansraj Gupta's designation was raised to that of a Lecturer.

When he was working in the small provincial college at Hoshiarpur, Hansraj Gupta was encouraged by the Principal of that college to take up research seriously. The greatest difficulty faced by young Hansraj at that time was the lack of any possible infrastructure. The college library did not have any books which he needed nor did it subscribe to any mathematical journal. The Principal of the College, Prof. Bhatia again came to his rescue. He wrote to the University Library at Lahore and got him enlisted as a member. Occasionally, Hansraj Gupta made trips to Lahore to consult necessary books and journals and with administrative help from Prof. Bhatia was finally able to submit his Ph.D. thesis at the University of Panjab at Lahore in 1935. In 1936, he was awarded a Ph.D. degree in mathematics by Panjab University, Lahore. It was the first Ph.D. degree awarded by that university in its history of over fifty years. The thesis of Hansraj Gupta was titled "Contributions to the theory of numbers" and was examined by two most noted number theorists of that time. They were Profs. G. H. Hardy and J. E. Littlewood of the University of Cambridge, England. They were impressed by the contents of the thesis and in their reports mentioned that they found it to be well above the standard of the average Ph.D. thesis. Another interesting point needs to be mentioned. In those days, typing facilities, especially typing tables, were neither good nor readily available. Hansraj Gupta's handwriting was very good almost like a typed script. So he had been granted special permission to submit a hand-written thesis.

In this connection, it would be relevant to mention that Dr. Hansraj Gupta had been inspired to carry out research on number theory by Pandit Hemraj who had been the Principal of the Dyal Singh College at Lahore. As already mentioned, Hansraj was a student there during his postgraduate studies. In the preface to the monograph titled *Symmetric Functions in the Theory of Integral Numbers*,[87] Dr. Hansraj Gupta stated:

---

[87]Lucknow University Studies Series, (1940).

... my revered teacher the Late Pandit Hemraj, who passed away on the 12th November 1938, instilled a lot of enthusiasm for the subject among his pupils and himself made valuable contributions to the subject. The little that I have done by way of research is in no small measure due to the encouragement I got from him....

Dr. Hansraj Gupta continued working and teaching at the Government College at Hoshiarpur till 1954. Though he had obtained his Ph.D. degree in 1936, he got promoted to the gazetted rank of the Panjab Educational Service, Class II, only in 1945. He was selected for Class I in the Panjab Educational Service in 1954. After the partition of the Indian subcontinent and the creation of Pakistan in 1947, the University at Lahore was lost to India. The Government College at Hoshiarpur was taken over by Panjab University and along with several other departments, the Mathematics department of the university was accommodated there. Dr. Hansraj Gupta was made a professor and put in charge of the new Department of Mathematics at the university. In 1958, the Department of Mathematics was shifted to the new university premises at Chandigarh. Professor Hansraj Gupta left Hoshiarpur after a stay of 30 long years and got settled at Chandigarh. The other faculty members of the Department of Mathematics at the new Panjab University were Dr. R. P. Bambah, Mr. T. P. Srinivasan, and Dr. I. S. Luthar. The new department made rapid strides and made a mark in the academic world. Under the benign guidance of Prof. Hansraj Gupta and due to the excellent work done by Dr. R. P. Bambah and the other colleagues, the University Grants Commission gave it the status of a Centre of Advanced Study in mathematics under its scheme of Centres of Excellence. In January 1964, Prof. Hansraj Gupta was made the first Director of the Centre. He finally retired from the university service in 1966. As a mark of recognition of his contributions to mathematical teaching, research, and advancement of the subject, he was made the Honorary Professor of mathematics at the University of Panjab.

The research contributions of Prof. Hansraj Gupta with particular emphasis on number theory will now be discussed. He made significant contributions in the area of number theory in particular. His substantial contributions are in partition theory, Diophantine equations, and elementary number theory. His contributions to combinatorics and discrete mathematics do not fall in the category of number theory. So they will not be discussed much. According to experts in the subject who knew Prof. Hansraj Gupta's work well, in the theory of partitions, he studied the generating functions, recurrence relations, congruences, identities, exact formulae, and asymptotics of various partition numbers. His work is remarkable for its simplicity, ingenuity, and elegance.

Tables of Partitions and Partitions

Professor Hansraj Gupta is internationally acclaimed for his researches in partition theory and related tables. But before the commencement of the serious discussion, it would be interesting to relate the events that led him to take up research in this area. The event has been related by Prof. Gupta himself. According to him, sometime in 1929 and 1930, a student approached him for solving a problem from a textbook on algebra and stated:

In how many ways can four mangoes be distributed among four persons when there is no restriction as to the number of mangoes any of them may receive?.

The student had earlier gone to another teacher who had taken the mangoes to be all different and had obtained the answer to be $4^4 = 256$. As the answer did not tally with the answer given in the book, the student had come to Prof. Hansraj Gupta. Professor Gupta considered the mangoes to be all alike and solved the problem accordingly. The answer was the same as was given in the book. The student became very happy that the problem was solved correctly. But according to Prof. Hansraj Gupta:

> Here the student's problem ended but mine had begun. I had noticed that what I had done was to partition 4 into at most four parts. I asked myself 'How many partitions will a given number n have into a given number of parts?' This is what I started investigating.

So that is how his journey into the realm of "partition theory" began.

Unfortunately, Hansraj Gupta had spent most of his working life in small places like Hoshiarpur and in small colleges where there were no libraries of any standing. He did not have any colleagues with whom to discuss mathematics. So he was quite unaware of the developments that took place in the area of research. Under his own admission, he did not even know that the problem he took up for investigation was not at all new. He was quite unaware of the fact that Srinivasa Ramanujan had already calculated and constructed a table giving the number of unrestricted partitions of $n \leq 200$. He had not seen Ramanujan's *Collected Papers* nor did he have any knowledge about the pentagonal number identity of Euler. He was passionately interested in partitions and their congruence properties. He made contributions in this area throughout his life. His first publication in this area was titled "A table of partitions".[88] Later, he further extended the tables and was a part of his thesis. As universally accepted, his most outstanding contribution is his "Tables of partitions". In 1939, they were first published as collected from his thesis as *Tables of Partitions*.[89] Later in 1958, more extended versions of these tables were published by the Royal Society of London.[90] These tables have been of extensive use not only in mathematics but also in diverse branches of science such as statistical mechanics and computer science. Dr. Hansraj Gupta had gained international fame for these *Tables*. The meticulous care that he took and his clear print-like handwriting made them error-free. S. Ramanujan and P. A. MacMahon (1854–1929) had independently calculated the values of $p(n)$, the number of unrestricted partitions of n, up to $n = 200$. Using many labor-saving devices and results, Hansraj Gupta had extended the table up to $n = 300$. In this context, it may be noted that in 1919, S. Ramanujan had proved that the partition function $p(x)$ satisfies the following congruences:

$$p(5m + 4) \equiv 0(\bmod 5),$$

---

[88] *Proceedings of the London Mathematical Society*, 39, (1935), 142–149.

[89] Indian Mathematical Society, Presidency College, Madras, (1939).

[90] *Tables of Partitions,* Royal Society Mathematical Tables, University Press, Cambridge, (1958), reprinted (1962).

$$p(7m + 5) \equiv 0(\bmod 7),$$
$$p(11m + 6) \equiv 0(\bmod 11).$$

He further proved that

$$p(25m + 24) \equiv 0\left(\bmod 5^2\right),$$
$$p(49m + 47) \equiv 0\left(\bmod 7^2\right)$$
$$p(121m + 116) \equiv 0\left(\bmod 11^2\right).$$

He finally generalized these results and conjectured that
if $\delta = 5^a 7^b 11^c$ and $24\lambda \equiv 1 \pmod \delta$,then $p(m\delta + \lambda) \equiv 0 \pmod \delta$.

for every m. This conjecture of Ramanujan led to a lot of research work in this area. As already mentioned earlier, Dr. Hansraj Gupta had extended the partition tables up to $n = 300$. He found that $p(243) = 133, 978, 259, 344, 888$, which is not divisible by 7^3. In his paper titled "Congruence properties of partitions",[91] S. Chowla observed that since $24 \cdot 243 \equiv 1 \pmod{7^3}$, this contradicted Ramanujan's conjecture. So in 1935, using Hansraj Gupta's tables, S. Chowla disproved one of Ramanujan's conjectures. The conjecture was later modified by G. L. Watson (1909–1988) and A. O. L. Atkin (1925–2008). They proved that if $24n - 1 \equiv 0 \pmod{5^a 7^b 11^c}$, where a, b, c are non-negative integers, then $p(n) \equiv 0 \pmod{5^a 7^d 11^c}$, where $d = [(b + c)/2]$. Apart from this work on partitions, Prof. Hansraj Gupta wrote about 70 research papers on problems related to partitions. In the process, he obtained a variety of results, solved a number of conjectures, and also posed many a problem. In this context, it would be interesting to discuss Prof. Hansraj Gupta's role in solving the Churchhouse conjecture concerning binary partitions. In 1969, R. F. Churchhouse studied the function $b(n)$ giving the number of partitions of n into powers of 2. He conjectured that:

The largest power of 2 which divides $b(2^{k+2}n) - b(2^k n)$ for $k \geq 1$ and n odd is $[(3k + 4)/2]$, where $b(n)$ is the number of partitions of n into powers of 2.

In 1970, O. Rodseth first proved the conjecture in 1970. Professor Hansraj Gupta gave three different proofs to the said conjecture. He published the first one in 1971 and simpler proofs in 1972 and 1976. He obtained generalizations of the famous Rogers–Ramanujan identities. During the late fifties and early sixties of the twentieth century, he wrote several papers on "partitions of j-partite numbers". His last paper on partitions was titled "Diophantine equations in partitions".[92]

Professor H. Gupta's work on number theory and his tables on partitions have been referred to in many standard textbooks, to name a few:

(i)   *Encyclopaedia Britannica*, Vol. 16, p. 604 (the article: "Theory of numbers");
(ii)   G. H. Hardy and E. M. Wright. *Theory of Numbers;*

---

[91] *Journal London Math. Soc.*, 9, (1934), 247.
[92] Section 2, (HG. 175).

(iii) D. H. Lehmar: *Guide to Mathematical Tables;*
(iv) G. H. Hardy. *Ramanujan;*
(v) *Hand-Book of Mathematical Tables.* National Bureau of Standards, U.S.A.

The National Bureau of Standards invited Prof. Hansraj Gupta to write a survey article on "Theory of partitions" for publication in their journal. Professor H. Gupta worked hard and wrote a remarkable article, which was published in 1970. It is considered to be one of the finest survey articles on the subject. It may be relevant to mention here that earlier in December 1963, Prof. H. Gupta had delivered a presidential address (technical) at the 29th Annual Conference of the Indian Mathematical Society at Madras. There he spoke on "Partitions: a survey". His address was later published as a paper in *mathematics Student*.[93] There he discussed in detail about different types of partitions, such as "restricted partitions", "decompositions", "perfect partitions", "cyclic partitions", "generating functions", "recursion formulae", "congruences", "Ramanujan's conjecture", "Newman's conjecture", "rank of a partition", "Hardy–Ramanujan–Rademacher series for $p(n)$", and "partitions into powers of primes".

In this context, it may be mentioned that the paper which he had co-authored with S. D. Chowla and F. C. Auluck and titled "On the maximum number of partitions of n into k parts"[94] generated a lot of interest among noted number theorists such as Erdös and Lehmar, and they did further research in this area.

Apart from partition theory and tables, Hansraj Gupta also had lots of interest in arithmetical functions and particularly in their congruence properties. He made significant contributions to Ramanujan's function $\tau(n)$ and the arithmetical functions such as $\sigma(n)$, $\sigma_k(n)$, and Euler's totient function $\varphi(n)$.

Professor H. Gupta was like many other Indian mathematicians enamored by the achievements of Srinivasa Ramanujan. He worked on topics which were brought to the limelight by Ramanujan. Ramanujan's function $\tau(n)$ was one such topic. In his paper titled "Congruence properties of $\tau(n)$",[95] he extended Ramanujan's table of values of function $\tau(n)$ for values of n up to 130. In a subsequent paper titled "A table of values of $\tau(n)$",[96] Prof. H. Gupta further extended the table up to $n = 400$. This was a useful addition. In his paper titled "Congruence properties of Ramanujan's function $\tau(n)$",[97] written in collaboration with R. P. Bambah, S. Chowla, and D. B. Lahiri, relationships between $\tau(n)$ and $\sigma(n)$ under different conditions were established. In another paper written in collaboration with R. P. Bambah and S. Chowla entitled "A Congruence Property of Ramanujan's Function $\tau(n)$",[98] it was proved that

$$\text{if } (n, 2) = 1, \text{ then } \tau(n) \equiv \sigma(n)(\bmod 8)$$

---

[93] 32, (1964), Appendix 1–19.
[94] Section 2, (HG. 46).
[95] Section 2, (HG. 53).
[96] Section 2, (HG. 65).
[97] Section 2, (HG. 62).
[98] Section 2, (HG. 63).

and

$$\text{if } 2|n, \quad \text{then } \tau(n) \equiv 0(\bmod 8).$$

In yet another paper connected to Ramanujan's function $\tau(n)$ and titled "The vanishing of Ramanujan's function $\tau(n)$",[99] Prof. H. Gupta could not verify a result because of the inaccessibility of a table of primes. That reflects the kind of difficulties he had to face while conducting his research work.

Representations of Primes by Quadratic Forms
In 1960, in collaboration with J. C. P. Miller, M. S. Cheema, A. Mehta, and O. P. Gupta, Prof. Hansraj Gupta published a monograph titled *Representation of Primes by Quadratic Forms.*[100] They published the solutions of the equation $x^2 + Dy^2 = p$ or $2p$ for $D = 5, 6, 10, 13$, and primes $p \leq 10^5$.

It may be noted that Prof. Hansraj Gupta was also responsible for publishing tables of values of Liouville's function $L(t)$. In his paper titled "A Table of Values of Liouville's Function $L(t)$",[101] he stated that if p denoted a prime ≥ 2, then for positive integral values of t, Liouville's functions $\lambda(t)$, and $L(t)$ were defined by the relations $\lambda(0) = 0, \lambda(1) = 1, \lambda(pt) = -\lambda(t)$, and $L(t) = \lambda(1) + \lambda(2) + \lambda(3) + \cdots + \lambda(t)$. Thus $\lambda(475) = -\lambda(95) = \lambda(19) = -\lambda(1) = -1$. In 1919, Polya conjectured that for values of $t \geq 2, L(t) \leq 0$. He verified this conjecture for values of t up to 1500. In 1940, at Dr. Chowla's suggestion, Dr. Hansraj Gupta computed a table giving the values of $\lambda(t)$ and $L(t)$ for values of t up to 20,000 and he also found Polya's conjecture to be true to that extent. With no modern computational facilities, it is indeed a remarkable piece of work.

Symmetric Functions
Symmetric functions $G(n, r)$ are also called *Stirling numbers of the first kind.* During 1931–1935, Hansraj Gupta made a thorough study of these functions and the results obtained formed the first part that is part one of his Ph.D. thesis. Later, these papers were published in the form of a book titled *Symmetric Functions in the Theory of Integral Numbers.*[102] The numbers $G(n, r)$ are expressed by the generating function

$$(x + 1), (x + 2), \ldots, (x + n) = \sum G(n, r)x^{n-r} \text{ with } G(n, 0) = 1.$$

He expressed this function in terms of combinatory functions, Bernoulli numbers, difference operators, etc., and obtained recurrence formulae. The symmetric functions of "self-contained balanced sets" were also studied by him. The properties

[99]Section 2, (HG. 67).

[100]Royal Society Mathematical Tables, University Press, Cambridge, (1960).

[101]Section 2, (HG. 73).

[102]Lucknow University Studies, Allahabad Law Journal Press, Allahabad, India, No. 14, (1940).

deduced by him showed that the theorems of Lagrange, Wilson, Fermat, Gauss, and many others follow as special cases.

Diophantine Equations

Professor Hansraj Gupta derived $=$ a method for writing the general solution of the Diophantine equation $x_1, x_2, \ldots, x_n = y_1, y_2, \ldots, y_m$ in terms of mn parameters. He used this method to solve many other Diophantine equations of the type:

$$ax^2 + y^2 = z^2 = x_1 x_2^2 = y_1 y_2 = z_1 z_2, \quad x^n = y_1, y_2, \ldots, y_m.$$

He proved some results which were related to the Tarry–Escott problem. It may be noted that his two publications given below are interesting:

- "On $N_q(r)$ in the Tarry–Escott problem" [Sect. 2.2, (HG. 68)].
- "A solution of the Tarry–Escott Problem of degree r" [Sect. 2.2, (HG. 70)].

Professor Hansraj Gupta also wrote a number of papers on various other topics in elementary number theory. He included many of them in his book titled *Selected Topics in Number Theory*.[103] They are written in simple language and are very useful for students and research workers on number theory.

Professor H. Gupta has made notable contributions to combinatorics. His paper written in collaboration with H. Anand and V. C. Dumir and titled "A combinatorial distribution problem"[104] is especially important, because, in this paper they made a certain mathematical conjecture. In 1974, using the powerful theory of Cohen–Macaulay rings and many more generalized results, *R.* Stanley proved the conjecture. These methods were later generalized and are called *Cohen–Macaulay partially ordered sets*. This has applications in number theory.

In his long active academic career of more than 60 years, Prof. Hansraj Gupta has published 190 research papers in national and international journals. He has also published 6 books and monographs. He has collaborated with 16 mathematicians. They are Harsh Anand, F. C. Auluck, R. P. Bambah, G. Bhattacharya, M. S. Cheema, S. D. Chowla, V. C. Dumir, Paul Erdös, O. P. Gupta, S. P. Khare, D. B. Lahiri, G. Baikunth Nath, P. A. B. Pleasants, Kuldip Singh, Seshadri Srinivasan, and A. M. Vaidya.

In recognition of his mathematical contributions of high order, he received several awards and honors. He was elected Fellow of the National Institute of Sciences, India (now known as Indian National Science Academy) and a Fellow of the National Academy of Sciences, Allahabad. In 1963, he was elected President of the Indian Mathematical Society. In 1979, the Mathematical Association of India conferred the "Distinguished Service Award" to him.

Two comments by two eminent mathematicians will suitably describe the great number theorist Prof. Hansraj Gupta who not only did great work himself but inspired generations of young students and research workers. G. E. Andrews commented:

---

[103] Abacus Press, India, (1980).
[104] Section 2, (HG. 105).

Gupta's mathematical career was, at many stages, isolated from other mathematicians with similar interests, and, at times, he had limited access to journals. Consequently, some of his work turned out to be rediscovery. However, even when that was the case, there was always a special 'Gupta-esque' twist.

Apart from his extensive work on tables, he provided a most elegant and elementary proof of the Erdös–Lehmar theorem on the asymptotics of the number of partitions of n into m parts. He gave a lovely, elementary proof of the Churchhouse conjecture on binary partitions. He also successfully attacked some of the hardest partition problems posed by Erdös.

His one-time colleague, Prof. I. B. Passi wrote:

The fact that Professor Gupta was the first person to be awarded Ph.D. degree by Panjab University, Lahore, and had produced high quality mathematics sitting at an isolated place like Hoshiarpur was a great source of inspiration and motivation for all students.

Another iconic number theorist of the Panjab School will be the subject of discussion next. He is Prof. R. P. Bambah. It has already been mentioned that S. Chowla as the founding father of the Panjab School of number theory trained up many noted mathematicians. He practically brought Hansraj Gupta to the foreground of the world of number theory. But one may claim without any exaggeration that Prof. Chowla's greatest contribution was in inspiring R. P. Bambah to carry out research on topics related to number theory.

2.2.3   Ram Prakash Bambah

R. P. Bambah (Fig. 2.5) was born on September 30, 1925, at Jammu. After completing his school education very successfully in 1939, he was admitted to the Government College at Lahore in undivided India (now in Pakistan) for pursuing further studies. He had an outstanding and congratulatory academic career. At the University of Panjab at Lahore, he stood first in the F.A. (equivalent to Intermediate), B.A., B.A. (Honors), and M.A. examinations. In the M. A. examination, he set an unbeatable record by scoring the full marks of 600 out of 600.

His exceptional talent in mathematics was noticed right from the beginning of his college days at Lahore. Eminent number theorist S. Chowla was among Bambah's teachers at the Government College, Lahore. During his student days, Bambah was highly impressed and motivated by Prof. S. Chowla. Later, once Bambah commented:

Professor Chowla was a real successor to Srinivasa Ramanujan, the greatest mathematician of Indian origin.

After completing his postgraduate studies in 1946, R. P. Bambah decided to carry out research on number theory under the supervision of S. Chowla. The first paper of Bambah published in collaboration with S. Chowla dealt with integer roots of a unit matrix. The paper was titled "On integer roots of a unit matrix".[105] The next paper

---

[105] Section 2, (RPB. 1).

Fig. 2.5 R.P. Bambah
(1925)

was published by Bambah independently and was titled "On Complete primitive residue sets".[106] Both these papers belong to elementary number theory.

Professor S. Chowla introduced R. P. Bambah to the work of the legendary Indian number theorist and mathematician Srinivasa Ramanujan. Inspired by him, Bambah took up studies on Ramanujan's τ-function. Ramanujan had introduced and defined τ-function as

$$\sum_{n=1}^{\infty} \tau(n)x^n = x[(1-x)(1-x^2)\cdots]^{24}$$

This function is known as *Ramanujan's τ-function*. From the above expression, it is clear that $\tau(n)$ is the coefficient of x^n in the polynomial which results when terms up to $(1-x^n)$ in the product on the right-hand side are taken.

Ramanujan did a lot of work on this function. He calculated the values of $\tau(n)$ like $\tau(1) = 1$, $\tau(2) = -2$, $\tau(3) = 252$, $\tau(4) = -1472$, ..., etc. He thus prepared tables for τ-functions. He also made many important conjectures on $\tau(n)$. Many years after Ramanujan's untimely demise, during 1936–1940, Prof. G. H. Hardy delivered a series of lectures on Ramanujan's research contributions. The 10th lecture delivered by Hardy was devoted entirely to τ-functions. In the said lecture, Hardy remarked:

I shall devote this lecture to a more intensive study of some of properties of Ramanujan's functions $\tau(n)$, which are very remarkable and still imperfectly understood. We may seem to be straying into one of the backwaters of mathematics. But the genesis of $\tau(n)$ as a coefficient in so fundamental a function compels one to treat it with respect.

---

[106] Section 2, (RPB. 2).

Bambah was highly inspired by these talks and started his research in this area. In 1946, in collaboration with S. Chowla, he published a series of papers related to Ramanujan's function. They are listed below:

- (With Chowla, S.) "A congruence property of Ramanujan function $\tau(n)$" [Sect. 2.2, (RPB. 4)]
- (With Chowla, S.) "On a function of Ramanujan" [Sect. 2.2, (RPB. 6)]
- (With Chowla, S.) "Some new congruence properties of Ramanujan's function $\tau(n)$" [Sect. 2.2, (RPB. 8)].

In 1947, R. P. Bambah collaborated with S. Chowla, Hansraj Gupta, and D. B. Lahiri at different times in connection with different problems related to Ramanujan's function. The result was the publication of the following seven papers. They are listed below:

- (With Chowla, S.) "A note on Ramanujan's function $\tau(n)$" [Sect. 2.2, (RPB. 9)]
- "Ramanujan's function: a congruence property" [Sect. 2.2, (RPB. 10)]
- (With S. Chowla and H. Gupta) "A congruence property of Ramanujan's function $\tau(n)$" [Sect. 2.2, (RPB. 11)]
- (With S. Chowla) "A new congruence property of Ramanujan's function $\tau(n)$" [Sect. 2.2, (RPB, 12)]
- (With S. Chowla, H. Gupta, and D. B. Lahiri) "Congruence properties of Ramanujan's function" [Sect. 2.2, (RPB. 13)]
- (With Chowla, S.) "Congruence properties of Ramanujan's function $\tau(n)$" [Sect. 2.2, (RPB. 14)]
- (With S. Chowla) "The residue of Ramanujan's function $\tau(n)$ to the modulus 2^8" [Sect. 2.2, (RPB. 17)].

The notable fact is between 1946 and 1947, Bambah independently or in collaboration with Chowla and some other number theorists published 10 research papers on topics related to Ramanujan's function $\tau(n)$. This clearly indicates how Ramanujan had inspired R. P. Bambah in the early years of his research career. In these two years, 1946 and 1947, Bambah and Chowla proved congruences for $\tau(n)$ modulo various powers of 2, 3, 5, and 7.

In 1946, S. Chowla had sent his earliest papers to Prof. G. H. Hardy in England. Professor Hardy had edited and combined these papers and got them published in the *Journal of the London Mathematical Society* as "Two congruence properties of Ramanujan's function $\tau(n)$".[107] Years later, Prof. R. P. Bambah had commented:

Because of the later work of many powerful mathematicians, Ramanujan's $\tau(n)$ function occupies a very central position in mainstream mathematics...because of its connection with modular forms, elliptic curves and so on, which play an important role in modern mathematics and in fact, have been crucial in the solution of Fermat's last theorem.

---

[107] Section 2, (RPB. 3).

When Bambah was working under Prof. Chowla's supervision, in 1947, they made an interesting conjecture in order to prove the existence of a constant C such that for $x \geq 1$, there is at least one integer between x and $x + Cx^{1/4}$, which can be expressed as a sum of two squares. They published their findings in the paper titled "On numbers which can be expressed as a sum of two squares".[108]

At that time, they had doubts about the reliability of the result. Later, many mathematicians like K. N. Majumdar (1950), S. Uchiyama (1965), P. H. Diananda (1966), and L. J. Mordell (1969) have worked on the value of C and tried to generalize the result. But not much improvement could be done to the result as established by Bambah and Chowla. In 1971, another British mathematician named C. Hooley proved some related results.

Due to political instability and the impending partition of India, in 1947, R. P. Bambah left Lahore and had a brief stint at the University of Delhi. In 1948, Bambah obtained a prestigious two-year scholarship and left for the University of Cambrigde in England. There he joined as a research student under the guidance of L. J. Mordell. Mordell was a well-known number theorist noted for solving Ramanujan's conjectures and many other important problems in number theory. In January 1949, he suggested a rather difficult problem to R. P. Bambah. The problem was related to numbers. Before discussing Bambah's contributions in this area, a little introduction is necessary.

Mathematical greats like C. F. Gauss (1777–1855) and J. L. Lagrange (1736–1813) had used geometrical methods for solving arithmetical problems. H. Minkowski (1864–1909) was a noted German mathematician who observed that geometrical methods when used in the case of some inequalities led to simple proofs of the existence of solutions in terms of integers (not all zeros). He also named it as *geometry of numbers*. The advantage of the method is that one has to only verify if the n-dimensional body defined by the inequality has certain properties. Soon "geometry of numbers" was developed into a major new branch of number theory. Many mathematicians started investigating new problems arising out of this new area. As for R. P. Bambah, he was at the right place, in the midst of great number theorists who were actively involved in conducting exciting research in the "golden age" of the geometry of numbers.

To explain Bambah's research work in this field, a few basic terminologies need to be introduced. In this context, a quotation from well-known number theorists R. J. Hans-Gill and S. K. Khanduja would be useful. According to them:

> We consider the n-dimensional Euclidean space R^n. A lattice is said to be admissible for a set S in R^n, if it has no point other than the origin in the interior of S. The critical determinant of S is the infimum of the determinants of lattices admissible for S. A fundamental problem in the geometry of numbers is to determine the critical determinant of a given set. For convex sets centered at the origin in the plane, general results are known, which at least theoretically give the critical determinant. But for non-convex sets, problems are more difficult. Techniques have to be found and details worked out for different regions. The difficulty increases in higher dimensions both for convex and non-convex sets. For subsets S, T of R^n, (S, T) is called a *packing* if the sets obtained by translating S through points of T are non-overlapping.

---

[108] Section 2, (RPB. 15).

This concept is related to the concept of admissibility. Also (S, T) is called a *covering* if each point of R^n lies in some translate of S through a point of T.

As stated earlier, in January 1949, Bambah took up the problem suggested by Mordell and within four months in April of the same year, he was successful in solving the problem. At first, Mordell was not ready to accept that the solution was correct, because he had thought that the problem was very difficult to be solved in such a short time. However, when J. W. A. Cassels confirmed the correctness of the solution, Mordell examined Bambah's solution seriously and was satisfied. Bambah soon submitted his Ph.D. thesis and in 1950, the University of Cambridge awarded him the coveted Ph.D. degree. In his thesis entitled *Some Results in the Geometry of Numbers,* Bambah successfully developed a technique for finding out the critical determinant of non-convex star regions with hexagonal symmetry. He extended some of the results which had been established earlier by his supervisor L. J. Mordell. In 1951, the following 4 papers were published from Bambah's thesis. They are listed here:

- "On the geometry of numbers of non-convex star regions with hexagonal symmetry" [Sect. 2.2, (RPB. 19)].
- "Non-homogeneous binary cubic forms" [Sect. 2.2, (RPB. 20)].
- "Non-homogeneous binary quadratic forms: Two Theorems of Varnavides" [Sect. 2.2, (RPB. 21)].
- "Non-homogeneous binary quadratic forms II: The second minimum of $(x + x_0)^2 - 7(y + y_0)^2$" [Sect. 2.2, (RPB. 22)].

Bambah spent the remaining one year of his scholarship working in collaboration with H. Davenport at the University College of London. During 1950–1951, he in collaboration with H. Davenport, C. A. Rogers, and K. F. Roth wrote the following three research papers, dealing with lattice coverings:

- (With K. F. Roth) "A note on lattice coverings" [Sect. 2.2, (RPB. 24)].
- (With H. Davenport) "The covering of n-dimensional space by spheres" [Sect. 2.2, (RPB. 25)].
- (With C. A. Rogers) "Covering the plane with convex sets" [Sect. 2.2, (RPB. 23)].

Dr. R. P. Bambah was elected as Fellow of St. John's College, Cambridge, from 1952 to 1955. In 1951, Bambah returned to India and joined as a research fellow at the National Institute of Science of India (now, known as Indian National Science Academy) in New Delhi. During his tenure at the National Institute, Bambah developed the theory of coverings. In "geometry of numbers", historical evidences show that more mathematicians were drawn toward the theory of packing than the concept of covering.

In 1945, soon after completing his M. A. examination, R. P. Bambah at the suggestion of his teacher Prof. S. Chowla went to meet Dr. Hansraj Gupta with the intention of finding some employment. At that time, Dr. Gupta was teaching at the Government College, Hoshiarpur. That was the first time Bambah met Dr. Gupta. Though

Bambah's results were not yet declared, the Principal of the Government College agreed to Dr. Gupta's request and asked Bambah to start teaching from the very next day. Dr. Gupta was extremely cordial and helpful and Bambah spent the entire three months of his temporary teaching assignment as a very welcome house guest of Dr. Gupta and his family. This was responsible for a very close relationship between Dr. Hansraj Gupta and Dr. Bambah.

In April 1952, Dr. Bambah was selected as a reader at Panjab University Department of Mathematics based at Hoshiarpur. It was then that Dr. Bambah came to work very closely with Prof. Hansraj Gupta. He had very fond memories of those days. When Dr. Bambah was selected at Panjab University, almost at the same time, he was offered a membership at the Institute of Advanced Studies at Princeton, U.S.A. Shortly after joining Panjab University, the then Vice-Chancellor granted him special leave for two years, so that Dr. Bambah could proceed to the U.S.A. and utilize the opportunities in Cambridge and Princeton. In 1954, Dr. Bambah returned to India and continued teaching at Panjab University. Dr. Gupta and Dr. Bambah with the help of a computation assistant started M. A. teaching at Hoshiarpur.

During 1957–1958, Dr. Bambah again went away from India and taught as a visiting professor for a year at the Notre Dame University. On his return, in 1958, he along with Prof. Hansraj Gupta and the computation assistant M. S. Cheema started the Department of Mathematics at the Panjab University at Hoshiarpur. But the untiring efforts of Bambah and Prof. Hansraj Gupta helped the department to develop fast. With help from the Vice-Chancellors, the University Grants Commission, Science Academies, and mathematicians from India and abroad, they created new posts. They selected suitable faculty members, got good students and that helped to create a vibrant and dynamic Department of Mathematics at the Panjab University. By the time the university shifted to Chandigarh, the department had acquired T. P. Srinivasan, I. S. Luthar, S. D. Chopra, M. L. Madan, Rajinder Singh, and many others. Professor Hansraj Gupta tried relentlessly and managed to get another professor's post sanctioned. So, in 1958 itself, R. P. Bambah became a professor at the Department of Mathematics. The Department grew rapidly under the benign guidance of Prof. Hansraj Gupta and the excellent work done by Prof. Bambah and other colleagues. The University Grants Commission granted it the status of a Centre of Advanced Study in mathematics under its scheme of Centres of Excellence. This department, with two excellent number theorists like Hansraj Gupta and R. P. Bambah, working and simultaneously supervising bright students, on problems related to number theory, became the bedrock of the famous Panjab School of Number Theory.

Going back to Prof. Bambah's research work, it may be noted that he did some fundamental work in the theory of coverings. A basic problem in this area is to determine lattice covering density θ_n of the n-dimensional sphere. In 1842, Dirichlet had obtained the result for a circle. More than a century later, in 1954, Bambah took up the work seriously and obtained results related to θ_3 for three-dimensional spheres. This gave rise to a new interest among mathematicians and led to further investigations. In 1956, E. S. Barnes and L. Few gave separate proofs. Subsequently, Bambah also obtained bounds on θ_4 and conjectured its value. He published the paper entitled "Lattice coverings with four dimensional spheres"[109] on this work. In 1963, B. N. Delone and S. S. Ryshkov and in 1967, T. J. Dickson gave proofs to Bambah's conjecture. Later in 1975, Ryshkov and E. P. Baranovski determined the lattice covering θ_5. Research work for the determination of lattice coverings in higher dimensions is still on. This clearly reflects the impact of Bambah's work in the international field of number theory.

It may be noted that as early as in 1952, Bambah and Davenport had successfully determined non-trivial lower bounds on θ_n. In the same year, Bambah and Roth were able to determine upper bounds for the lattice covering density of convex bodies which are symmetric about the coordinate plane. A lot of research was carried out in this area, the main target being to find out if the best lattice covering density of a given body was equal to its best general covering density. During the twentieth century, during the early fifties, Bambah in collaboration with C. A. Roth and in the late sixties in collaboration with Woods discovered new techniques to settle the proof for two-dimensional symmetric convex domains. They also showed that the result was true for a cylinder with a symmetric convex domain as base. These methods led to the study of finite coverings. This turned out to be a very important area of research.

In 1977, Bambah, V. C. Dumir, and R. J. Hans-Gill gave examples to show that there existed symmetric and asymmetric star domains in the plane for which the result mentioned earlier did not hold. The problem still remains unsolved for three-dimensional spheres.

Another important area where Prof. Bambah and his students have done a lot of work is on Minkowski's conjecture. In fact, the main thrust area of research on number theory of the Panjab school during the twentieth century has been on geometry of numbers comprising the concept of coverings and Minkowski's conjecture.

The conjecture of Minkowski may be stated as follows: Let $L = a_{i1}x_1 + a_{i2}x_2 + \cdots + a_{in}x_n$, $1 \le i \le n$, be n real linear forms in n variables $x_1, x_2, \ldots, x_n$ and having determinant $\Delta = \det(a_{ij}) \ne 0$. Then the following conjecture was made by H. Minkowski. He proposed: "For any real numbers $C_1, C_2, \ldots, C_n$, there exist integers $x_1, x_2, \ldots, x_n$ such that

$$|(L_1 + C_1)(L_2 + C_2)\ldots(L_n + C_n)| \le |\Delta/2^n$$

In 1899, Minkowski had proved the conjecture for $n = 2$. This conjecture has so far been proved for $n \le 5$. For $n \ge 3$, the proofs have been obtained by the following mathematicians listed below.

Remak (1921), Davenport (1939), and Birch and Swinnerton-Dye (1956), Narazuallev (1968); all for $n = 3$.

Dyson (1948), Skubenko (1973), and Bambah and Woods (1974); all for $n = 4$.

Bambah, R. P. and Woods, A. C.: "On a theorem of Dyson"; collection of articles dedicated to K. Mahler on the occasion of his seventieth birthday, [Sect. 2.2, (RPB. 53)] (No. 7, INSA).

---

[109] Section 2, (RPB. 29).

Skubenko (1973), Bambah and Woods (1980); all for $n = 5$.

Bambah, R. P. and Woods, A. C.: "Minkowski's conjecture for $n = 5$; *a theorem of Skubenko*" [Sect. 2.2, (RPB. 56)].

Considering the two publications of Bambah and Woods, it may be noted that they successfully proved the conjecture for values of $n = 3$ and $n = 5$ by using elementary techniques without taking help of strong tools from algebraic geometry. Proofs for the conjecture have been done for values of $n = 7$, 8, and 9 by R. J. Hans-Gill, Madhu Raka, Leetika, and others from the Panjab School working on number theory. But all these papers were published in the twenty-first century and so the details are not being considered here. Minkowski's conjecture for $n = 2$ has been interpreted as a result of non-homogeneous binary quadratic forms. Professor Bambah and his students R. J. Hans-Gill, V. C. Dumir, and others have done notable work in this area. Other number theorists of Panjab School such as V. K. Grover, R. Sehmi, Madhu Raka, and Urmila Rani have done extensive investigations in this and allied areas. Starting from 1979 till the late nineties of the twentieth century, they have published more than 25 research papers in this area. The paper entitled "Positive values of non-homogeneous quadratic forms of type (1,4)"[110] by Madhu Raka and Urmila Rani is supposed to be of significant importance.

Professor Madhu Raka of the Panjab School of Number Theory and her students and collaborators have made notable work in the area of Watson's conjecture. They have made important contributions in the field of quadratic forms which have led to a proof of Watson's conjecture, on non-homogeneous indefinite quadratic forms. In her Ph.D. thesis, Madhu Raka has determined the relevant constants $C_{n,\sigma}$ for $n = 5$ and all signatures. She also obtained $C_{n,\sigma}$ for signatures $+1$ and -1, $+2$ and -2, $+3$ and -3, $+4$ and -4, and all n. Several other mathematicians like Davenport, Birch, Watson and Dumir have worked towards the solution of this conjecture. Since the conjectured value of $C_{n,\sigma}$ depended on the class of σ modulo 8, this was an important and major contribution toward the proof which was completed by Dumir, Hans-Gill, and Woods in 1994. The work of Margulis (1987) on Oppenheim conjecture is also related to this problem in case of incommensurable forms.

Madhu Raka and her collaborators in a joint work have proved several results on the minimum value of $\Gamma_{n,\sigma}$ of positive values of non-homogeneous quadratic forms, which have also significantly contributed toward the proof of a conjecture of Bambah, Dumir, and Hans-Gill (1981). The only constant which could not be determined as yet is $\Gamma_{5,-3}$. However, the upper bound 12 for the same was obtained by Madhu Raka et al., in 1997, the expected value being 8.

Apart from topics related to number theory, Prof. Bambah has contributed to various other areas, such as integer matrices, polar reciprocal convex bodies, divided cells, transference theorems, lower bounds for minimum distance codes, convex bodies with covering property, maximal covering sets, and saturated systems. Professor R. P. Bambah has published singly and jointly around 70 research papers between 1946 and 2000 in various topics of number theory.

---

[110]Section 2, (MR. 17).

Professor R. P. Bambah was awarded the Sc. D. degree by Cambridge University in 1970. Since 1993, he is Emeritus Professor at Panjab University. Recently, he has been honored with the degree of Doctor of Science (Honoris Causa) by Panjab University.

Another important number theorist of the Panjab School is Prof. A. R. Rajwade. He has worked extensively in algebraic number theory, more specifically on the following three topics:

- Arithmetic on elliptic curves with complex multiplication
- Cyclotomy and roots of unity
- Sums and products of squares in fields and rings.

Professor Rajwade's co-workers have been late Dr. J. C. Parnami, late Dr. M. K. Agarwal along with his students Dr. Sitendra Pal Sharma, Dr. Dharambir Rishi, Dr. Surjit Singh, Dr. Budh Singh (settled in the USA), and Prof. S. A. Katre.

Professor Rajwade's work on elliptic curves has been on curves with complex multiplication by cube roots of unity, $\sqrt{-2}$, $\sqrt{-7}$, and $\sqrt{-19}$. He and his collaborators have verified Swinnerton-Dyer conjectures for these curves. The related publications are given in Sect. 2.3 of his list of publications.

In cyclotomy and roots of unity, Prof. Rajwade and his co-workers have completely solved the cyclotomic problem for a general prime p. The related publications are in Sect. 2.3. The paper numbered 15 of this section is particularly important. In the third topic mentioned above, apart from a number of other results proved by them, the most interesting result is given below:

Theorem: Let $F = Q(\sqrt{-m})$ be the quadratic field with m square-free and $\equiv 7$ mod 8. The minimal s(called the 4th power Stufe of F) for which the equation $-1 = a_1^4 + a_2^4 + \cdots + a_s^4$ is solvable in F, is 15.

The list of research publications of A. R. Rajwade (during the twentieth century) is given at the end of the report. Two noted women number theorists of Panjab School are Prof. R. J. Hans-Gill and Prof. S. K. Khanduja.

R. J. Hans-Gill obtained her B.A. (Honors) degree from the Government College for Girls, Ludhiana, Panjab. She obtained her M.A. from the same college with very good marks. She then joined the Department of Mathematics at Panjab University at Chandigarh and started her research career under the supervision of the famous number theorist Prof. R. P. Bambah. In 1962, Prof. Bambah decided to go to Ohio State University for a couple of years. Fortunately, his research scholars were also given scholarships there. So Hans-gill with support from her family moved over to the U.S.A. and finally obtained her Ph.D. degree from Ohio State University in 1965.

After brief teaching assignments at the same university and also at Madison, U.S.A., she finally returned to India and joined the Department of Mathematics of Panjab University as a reader. Subsequently, she became a professor there and served till her retirement in 2005. Her notable contributions are in number theory and geometry of numbers. Initially, she started working on extremal packing and covering sets and later on double packing and coverings. She obtained several results on non-homogeneous quadratic forms and was able to prove a conjecture made by Watson and also a conjecture of Dumir, Bambah, and Hans-Gill. Later with Dumir and Wilker, she made important contributions to the view obstruction problem of Schoenberg. In recent times, in collaboration with Madhu Raka and R. Sehmi, she has given a proof of a conjecture of Woods for inhomogeneous minima of positive definite quadratic forms in 7, 8 variables. She was thus able to complete the proof of a conjecture of Minkowski on the product of n non-homogeneous real linear forms in n variables for $n = 7, 8$. They have also obtained improved estimates on Minkowski conjecture. R. J. Hans-Gill has more than 50 research publications to her credit. Some of her more important publications are listed at the end of the report. Madhu Raka obtained her Ph.D. degree under the supervision of R. J. Hans-Gill.

Professor S. K. Khanduja is the other notable lady number theorist of the Panjab School. She studied and obtained her B.A. degree in mathematics from the Dev Samaj College for Girls in Panjab. She obtained her M.A. degree from the Department of Mathematics, Panjab University at Chandigarh. Thereafter, she started working on algebraic number theory under the guidance of Prof. I. S. Luthar at the Department of Mathematics of the Panjab University and obtained her Ph.D. degree from Panjab University in 1976. Her research works are mostly on theory of valuations, function field theory, and algebraic number theory. She has used theory of valuations to obtain the generalizations of the classical Schonemann–Eisenstein irreducibility criterion. She has used prolongations of valuations to generalize well-known Ehrenfeucht–Tverberg irreducibility criterion for difference polynomial. Some theorems of Dedekind have been extended by her to arbitrary valued fields. She has 65 research papers to her credit, but the list of her publications up to 1999 has been attached at the end of the report, because of the time frame of the present book. She has guided eight Ph.D. students.

Professor S. A. Katre is another important number theorist of India who is a product of the famous Panjab School. He did research work in algebraic number theory and cyclotomy. When he started research under the guidance of Prof. A. R. Rajwade from Panjab University, Chandigarh, there was a lot of work available in cyclotomy where a number of results about cyclotomic numbers, Jacobi sums, Jacobsthal sums, number of points in algebraic curves over finite fields, etc., were available up to a certain sign ambiguity or generator ambiguity, which used to be called a natural ambiguity in cyclotomy. S. A. Katre in his work showed how roots of unity in finite fields and expressions for the same in terms of solutions of certain Diophantine systems can be used to resolve these ambiguities. Stepping further on the work of Parnami, Agrawal, and Rajwade, he obtained the arithmetic characterization of Jacobi sums of prime order l, thereby giving a complete solution to the cyclotomic problem for order l. Correct results for roots of unity in finite fields, Euler's criterion for quintic residues, and Jacobsthal sums of order 4, 9, and prime order were obtained using these ideas.

S. A. Katre obtained his Ph.D. degree from Panjab University in 1984 for his thesis titled *Complete Solution of the Cyclotomic Problem for a Prime Modulus and Related Topics*. As already mentioned, his supervisor was Prof. A. R. Rajwade of the same University.

Dr. V. V. Acharya, a Ph.D. student of Prof. Katre, extended his work to cyclotomic numbers and Jacobi sums of order $2l$, l, an odd prime. Dr. Anuradha Narasimhan, another Ph.D. student of S. A. Katre, obtained, in the case of uniform cyclotomy, nice formulae for cyclotomic numbers, Jacobi sums, and Zeta functions of curves corresponding to l, $2l$, and l odd prime. S. A. Katre and Sangita Khule considered the problem of writing matrices over an order R in an algebraic number field as sums of kth powers. Answering a question of M. Newman, they showed that for $n \geq k$, every $n \times n$ matrix over R is a sum of kth powers if and only if $(k, disc. (R)) \equiv 1$.S. A. Katre also edited a book, *Cyclotomic Fields and Related Topics*.

2.3   Trends of Research on Number Theory in Bengal and Bihar

The states of Bengal and Bihar did not produce superlative number theorists like in the South Indian or the Panjab School. But these two states did produce some sincere mathematicians who did carry out research in the area of theory of numbers to the best of their ability.

2.3.1   Number Theorists of Bengal

In the twentieth century, the earliest known publication on number theory was by the well-known philosopher and mathematician Prof. Brojendra Nath Seal (or B. N. Seal) (1864–1938). He held the prestigious King George V Chair Professorship of Philosophy at the University of Calcutta. But he had a flair for mathematics in general. He published his paper entitled "The equation of digits; being an elementary application of a principle of numerical grouping to the solution of numerical equation".[111] In fact, much earlier in 1891, B. N. Seal had written "A memoir on the coefficient of number: a chapter on the theory of numbers". But since that was published in the nineteenth century, it is not within the purview of this book.

In 1919, H. Datta published a paper titled "On some properties of natural numbers".[112] After this, no notable publication on theory of numbers is noticed from eastern India during the next decade. S. C. Mitra incorporated a particular result on number theory (which was established by him) in his Ph.D. thesis in 1929. Three years later, this result was published as a paper entitled "On the proof of a result given by Ramanujan about the complex multiplication of elliptic function".[113] In this paper, Mitra has supplied the proof of a result about complex multiplication of elliptic functions, which the late S. Ramanujan had stated without a proof in his

---

[111] Section 4, (BNS. 1).

[112] Section 4, (HD. 1).

[113] Section 4, (SCM. 1).

famous paper titled "Modular Equations and Application to π".[114] The first investigation on a topic related to Ramanujan's work came from Bengal nine years after the death of the iconic mathematician.

Thereafter, three notable things took place, which helped the development of research on number theory in Bengal. Firstly, the famous number theorist T. Vijayaraghavan who has already been discussed earlier came away from Aligarh Muslim University and joined Dacca University in erstwhile East Bengal (present-day Bangladesh) in 1931. He stayed there till 1946. It was during this period that a lecturer in mathematics named D. P. Banerjee placed in an obscure college called A. M. College in Mymensingh in erstwhile East Bengal came in contact with Vijayaraghavan and got inspired and influenced by him. As a consequence, he started doing serious research on various topics connected to analytic number theory. D. P. Banerjee's first published paper was on Waring's problem. The paper was titled "On the solution of the 'easier' Waring problem".[115] Very briefly speaking, Waring's problem states that for every $k \geq 2$ there is a number $r \geq 1$ such that every natural number is a sum of at most rth powers. Subsequently in a series of two papers, Banerjee investigated the properties of Ramanujan's function $\tau(n)$. The papers published by him are listed below:

- "Congruence properties of Ramanujan's function $\tau(n)$" [Sect. 2.4, (DPB. 2)]
- "On the new congruence properties of the arithmetic function $\tau(n)$" [Sect. 2.4, (DPB. 3)].
- The next year, D. P. Banerjee published the paper titled "On the rational solutions of the Diophantine equation $ax^n - by^n = k$.[116] The next year in 1944, Banerjee a series of two papers titled:
- "On some formulae in analytic theory of numbers" [Sect. 2.4, (DPB. 5)]
- "On some formulae in analytic theory of numbers II" [Sect. 2.4, (DPB. 6)].

In these two papers—in continuation to certain formulae of analytic number theory, which were considered by Ramanujan, Estermann, and Hardy, as stated in the collected papers of Ramanujan[117]—Banerjee has added a few more interesting formulae to the list.

In his paper titled "On the application of the congruence property of Ramanujan's function to certain quaternary form",[118] Banerjee claims to have proved certain congruence properties of Ramanujan's function $\tau(n)$. It may be noted that in this paper, the author also considered new and interesting applications of the congruence properties to the possibility of finding applications to certain quaternary equations. The same year, Banerjee published a paper entitled "On a theorem in the theory of

---

[114]Section 1, (SR. 4).

[115]Section 4, (DPB. 1).

[116]Section 4, (DPB. 4).

[117]Edited by G. H. Hardy, P. V.Seshu Aiyer and B. M. Wilson: Cambridge University Press, Cambridge (1927).

[118]Section 4, (DPB. 7).

partitions".[119] Years later in 1964, he published another paper related to the theory
of partitions. The paper was titled "On some identities in the theory of partitions".[120]
This was probably his last publication related to the theory of numbers. In 1947, he
had published a paper related to the divisors of numbers.[121] In the ten researches
of D. P. Banerjee mentioned above, eight of them are connected to Ramanujan's
τ-function, partition functions, and Diophantine equations. These are areas where
Ramanujan had pioneered research in India. Clearly, like every other number theorist
of India during the twentieth century, D. P. Banerjee was also highly inspired and
motivated by Srinivasa Ramanujan.

The next important event that enriched the research on the theory of numbers in
Bengal relates to a mathematician named Deba Brata Lahiri (or D. B. Lahiri). He
was born on February 17, 1913, in Rangoon, the capital of erstwhile Burma (present-
day Myanmar). His family members were originally from East Bengal of undivided
India, but repeated natural disasters, poverty, and the onset of the First World War
forced the family to abandon their original homeland and migrate to Burma in search
of a new place to settle down. The family was extremely poor and due to precarious
financial conditions, young D. B. Lahiri had to depend upon stipends and scholarships
from the tender age of nine to carry on with his school education. He completed his
school final examination from a school in Rangoon and fared well. But being under-
aged, he could not qualify for a government scholarship. An Englishman and a
professor of mathematics, L. G. Owen, helped him to procure adequate funds. With
that financial help, Lahiri was able to continue his undergraduate studies. In 1933,
he passed his B.Sc. with Honors in mathematics and topped the list of successful
students and won a gold medal. Subsequently, with help from Mr. Owen and another
Englishman Dr. Lewis, Lahiri became well-trained in basic statistics. A research
paper on "Demography" which Lahiri co-authored with Mr. Lewis was published
in *Sankhya*, a journal of statistics. Professor P. C. Mahalanobis, who was the editor
of the said journal, must have noted the name of the young man. Lahiri came to
Calcutta from Rangoon in search of a job. After doing some odd jobs in different
places in and around Calcutta, Lahiri finally joined the Indian Statistical Institute in
Calcutta in the mid-forties of the twentieth century. While he was being trained on
specialized branches of applied statistics by Prof. P. C. Mahalanobis, Prof. S. N. Ray,
and others, the stalwart mathematician R. C. Bose came to know about his passion
for mathematics.

As a schoolboy in Rangoon, Lahiri had developed a fascination for the theory
of numbers. The news of the sudden and untimely death of Srinivasa Ramanujan
had left a lasting impact in Lahiri's mind. Since his young school days, D. B. Lahiri
became curious and eager to learn about Srinivasa Ramanujan and his contributions
to number theory.

While in the Indian Statistical Institute, R. C. Bose encouraged his love for number
theory and introduced him to the famous South Indian number theorist Dr. S. S. Pillai.

---

[119]Section 4, (DPB. 8).

[120]Section 4, (DPB. 11).

[121]Section 4, (DPB. 9).

At the initiative and persuasion of Prof. F. W. Levi, Dr. Pillai had joined the Pure mathematics Department of the University of Calcutta in 1942. D. B. Lahiri worked sincerely and very hard on the training he was receiving in applied statistics. But in his spare time, he assiduously carried out his private research on number theory.

As he was always fascinated by Ramanujan and his research on number theory, Lahiri's first publication on the subject was titled "On Ramanujan's function $\tau(n)$ and divisor function $\sigma_k(n) -$ I".[122] In this particular paper, Lahiri has developed a systematic method of studying certain congruence properties of the divisor function $\sigma_k(n)$, the sum of the kth powers of the divisors of n. In the same issue of the *Bulletin of the Calcutta Mathematical Society*, Lahiri published his paper entitled "On a type of series involving the partition function with applications to certain congruence relations".[123] He prepared this paper under the guidance of Prof. R. C. Bose. In this paper, Lahiri has discovered the recursive congruence properties of the partition function $p(n)$ and showed that the well-known Ramanujan's congruences

$$p(5m + 4) \equiv 0(\bmod 5)$$
$$p(7m + 5) \equiv 0(\bmod 7)$$
$$p(11m + 6) \equiv 0(\bmod 11)$$

can be deduced by induction method from the recursive congruences established by him. In 1947, Lahiri published two papers both dealing with Ramanujan's function $\tau(n)$. The first one was titled "On Ramanujan's function $\tau(n)$ and divisor function $\sigma_k(n) -$ II".[124] The second paper was written in collaboration with R. P. Bambah, S. Chowla, and H. Gupta and titled "Congruence properties of Ramanujan's function $\tau(n)$".[125] As is evident, Lahiri collaborated with leading Indian number theorists of his time. In 1948 and 1949, he published two papers dealing with non-Ramanujan congruence properties of the partition function. After a gap of more than 15 years, he again started publishing important research papers on topics of number theory. Between 1966 and 1971, he published as many as 11 research papers and they were all published in well-known national and international journals. Two of these papers deserve special mention. In the paper titled "Some congruences of the elementary divisor functions",[126] Lahiri has apart from carrying out studies on $\sigma_k(n)$, established two interesting theorems involving Euler's function $\varphi(n)$. In the other notable paper titled "Some restricted partition functions: congruences modulo 3",[127] the author has established some congruence relations with respect to modulo 3 for some restricted partition functions. He has also discussed unrestricted partition functions and stated two theorems involving them. Using the well-known "pentagonal number theorem"

---

[122] Section 4, (DBL. 1).

[123] Section 4, (DBL. 2).

[124] Section 4, (DBL. 3).

[125] Section 4, (DBL. 4).

[126] Section 4, (DBL. 13).

[127] Section 4, (DBL. 14).

of Euler, he has proved the theorems. D. B. Lahiri is an important number theorist from Bengal.

Apart from D. B. Lahiri, the renowned mathematician R. C. Bose who was a professor at the Indian Statistical Institute, Calcutta, contributed a couple of papers on number theory. His paper written in collaboration with S. Chowla and C. R. Rao, titled "On the integral order (mod p) of quadratics $x^2 + ax + b$ with applications to the construction of minimum functions for $GF(p^2)$, and to some number theory results",[128] is interesting. Here, the authors considered polynomials whose coefficients belonged to the ring of integers. In this paper, p always denotes an odd prime. Again if n is the least positive integer such that $x^n \equiv$ an integer (mod $p, x^2 + ax + b$), then n is called the *integral order* of $x^2 + ax + b$ (mod p). In this particular paper, the three authors have deduced a number of interesting theorems about the integral order. R. C. Bose published another paper titled "On the construction of affine difference sets"[129] in collaboration with S. Chowla.

The third important event that enriched the number theoretic researches in Bengal relates to Dr. S. S. Pillai of the University of Calcutta. As already mentioned, in 1942, Dr. Pillai joined the Pure Mathematics Department of the Calcutta University. Prior to that, S. S. Pillai and T. Vijayaraghavan had a long-standing friendship starting from their student days at Madras University. When Vijayaraghavan was in Oxford, Prof. G. H. Hardy discussed a problem which arose in the context of a famous paper of G. H. Hardy and S. Ramanujan titled "The normal number of prime factors of a number n".[130] This problem which Hardy posed to his students at Oxford came to be known as *Hardy's problem* and remained unsolved for many years. After returning to Madras from Oxford, when Vijayaraghavan met Pillai at Madras, he quietly passed on Hardy's problem to Pillai. Pillai in his paper titled "On the number of numbers which contain a fixed number of prime factors"[131] announced an important breakthrough to Hardy's problem. In 1943, as a faculty member of the University of Calcutta, Dr. Pillai started supervising a brilliant young student named L. G. Sathe. Sathe had obtained a research scholarship at Calcutta University. Pillai suggested the unsolved "Hardy's problem" to L. G. Sathe. He also gave him all the earlier manuscripts related to the problem. Noted number theorist Prof. R. Balasubramanian has nicely described Sathe's contributions as follows:

> ...In the course of less than two years, L. G. Sathe produced a monumentally complex induction argument, that ran into 134 printed pages when published, and that did much more than solve Hardy's problem. In particular, and for the first time, Sathe was able to show that Landau's asymptotic remained valid for $k < e \log \log x$. When finally in 1954, Sathe's result would supercede a theorem of Erdös, which appeared in 1948 and which solves Hardy's problem by obtaining Landau's asymptotic for k in an interval of length about $(\log \log x)^{1/2}$ around $\log \log x$. The sequence of events leading to publication of Sathe's work however remains somewhat opaque. It is clear, though, that Sathe submitted his work to the *Transactions of American Mathematical Society*...that Selberg's comments on this work

---

[128] Section 2, (SC. 116).

[129] Section 2, (SC. 126).

[130] Section 1, (SR. 16).

[131] MS, 14, (1929), 250–251.

was sought by the editors of the Transactions. Finally, however, Sathe's work appeared in the *Journal of the Indian Mathematical Society* ...in four parts, the first part in 1953 and the last in 1954.

Unfortunately, Sathe fell seriously ill soon after and became incapacitated. So, that cut short a brilliant mathematical talent, who might have contributed much more to the school of research of number theory in Bengal. Dr. Pillai also died of a plane crash in 1950 at the age of 49. These two major disasters practically stalled any further progress in research on the theory of numbers in Bengal. D. P. Banerjee, D. B. Lahiri, R. C. Bose, and L. G. Sathe stand out as important contributors in the area of number theory in Bengal. Their most productive years were between 1940 and 1970.

Bhaskar Bagchi worked for his Ph.D. degree and obtained the same from the Indian Statistical Institute, Calcutta, in 1981. The research work that he did was on probabilistic functional analysis. But this relates to Zeta function. Zeta function is intimately connected to number theory via the Euler product formula.

The Riemann zeta function assigns to every point of the complex plane, another point of this same plane (except at the point 1, where Zeta shoots off to infinity). Bagchi in his work defined the critical strip as the open strip in this plane bounded by the two vertical lines through ½ and 1 (actually, this is the right-half of what is usually called the *critical strip*). Riemann's famous paper revealed that most of the mystery of Zeta is concentrated in its behavior on that strip. In a very precise sense, this mystery encodes another mystery: that of the prime counting function.

In his Ph.D. thesis, titled "Statistical Behaviour and Universality Properties of the Riemann Zeta Function and Other Allied Dirichlet Series", Bagchi viewed the Zeta function as a single point in the infinite dimensional space consisting of all the complex differentiable functions on the critical strip. The vertical shifts of Zeta trace out the orbit of a particle moving in that abstract space. He showed that the asymptotic behavior of that orbit mimics the behavior of a related random function. Thus, in a precise sense, the mystery of Zeta is probabilistic in nature.

As a consequence of the result obtained by Bagchi, he could reinterpret and generalize the universality theorem of S. M. Voronin. For example, he could establish that the Zeta orbit came arbitrarily close to all points in the abstract space which corresponded to non-vanishing functions. That approach also allowed him to generalize the universality theorem for Riemann's zeta to a joint universality theorem for the *L*-functions introduced by Riemann's teacher Dirichlet. As another consequence, it was shown by him that the famous Riemann hypothesis (which is the statement that Zeta never assigned the value 0 anywhere on the critical strip, as defined earlier) amounted to a kind of almost periodicity of the Zeta orbit.

Bagchi's results on "joint universality theorem" and the "theorem relating Riemann hypothesis with almost periodicity" were published in *Mathematische Zeitschrift* and *Acta Mathematica Hungarica*. They are stated as below:

- "A joint universality theorem for Dirichlet *L*-functions", *Math. Zeitschrift*, 181, (1982), 319–334

- "Recurrence in topological dynamics and the Riemann hypothesis", *Acta Mathematica Hungarica,* 50, (1987), 227–240.

After completing his Ph.D., Bagchi moved away from number theory and did research in other areas such as finite geometries, coding theory, and combinatorial topology. He was probably the last noted number theorist produced from Bengal who carried out research during the twentieth century.

2.3.2 Number Theorists in Bihar

In Bihar, the first paper on number theory was published in the late forties of the twentieth century. It was authored by S. D. Upadhyay and P. N. Dasgupta; both were associated with Patna Science College. The paper titled "On a generalized continued fraction"[132] deals with continued fractions. In this paper, the authors state that a series of expressions for the numerator and denominator of any convergent of a continued fraction, by means of matrices, had been given by Milne-Thomson in 1933. The authors of this paper while working on a problem related to reciprocal differences in connection with interpolation formula had to consider continued fractions of a general type where coefficients themselves represent continued fractions.

A year later, guided by P. N. Dasgupta, Anunoy Chatterjee also from the Science College of Patna University published a paper titled "On a continued fraction of a general type".[133] It is stated by the author in this paper that Milne-Thomson in 1933 first exhibited a simple continued fraction in the form of a continued matrix product. Upadhyay and Dasgupta in their 1947 paper (already mentioned above) discussed continued fractions of a general type where various partial quotients were themselves continued fractions. In this paper, the author A. Chatterjee has discussed some of the formulae obtained by Upadhyay and Dasgupta and has extended the idea of continuing the type referred to by further assuming the partial quotients of the general type as being expressible themselves by continued fractions.

From the late sixties of the twentieth century, a number of mathematicians in Bihar contributed research papers on various topics of number theory. T. N. Sinha published a paper titled "Some systems of Diophantine equations of the Terry–Escott type".[134] M. R. Iyer, in 1969, published a series of three papers related to Fibonacci numbers and they were all published in *FQJ.* In 1971, T. N. Sinha again published a paper on integer solutions of equations of a special type and it was published in *The mathematics Student.* The other researchers from Bihar who published during the latter half of the twentieth century include S. A. N. Moorthy, P. D. Shukla, R. Tandon, S. N. Prasad, R. N. Lal, A. Ahmad, K. C. Prasad, R. N. Singh, D. N. Singh, S. N. Dubey, and A. Murthy. These mathematicians have worked on topics ranging

---

[132] BCMS, 39 (1–4), (1947), 65–70.

[133] BCMS, 40, (1948), 69–75.

[134] Section 4, (TNS. 1).

from prime numbers, Diophantine equations, and quadratic forms to divisibility of numbers.

2.4 TIFR School of Number Theory

The school of number theory that started at the Tata Institute of Fundamental Research (TIFR) in Bombay (present-day, Mumbai) from the middle of the twentieth century is an internationally renowned center of research in pure mathematics and theoretical physics. Many famous number theorists of India were trained and groomed at the mathematics school of TIFR. They have made remarkable contributions in this discipline. The concerned number theorists and their respective contributions will be discussed in detail. But at the very outset, it would be necessary to write about Prof. K. Chandrasekharan who was not only a leading Indian mathematician of the twentieth century, but was also responsible for building up the famous school of mathematics at the TIFR.

2.4.1 Komaravolu Chandrasekharan (1920–2017) and the Initial Years of the TIFR

Komaravolu Chandrasekharan (or K. Chandrasekharan) was born on November 21, 1920, at Machilipatnam in erstwhile Madras Presidency (present-day Andhra Pradesh) (Fig. 2.6). His father Rajaiah Chandrasekharan was the headmaster of a local school and his mother Padmakshamma a housewife.

K. Chandrasekharan had his schooling from the Bapatla village in the Guntur district of Andhra Pradesh. After completing his school education in 1940, he moved to the city of Madras (present-day Chennai) for pursuing higher studies. Through

Fig. 2.6 Komaravolu Chandrasekharan (1920–2017)

the famed Presidency College of the city, he obtained his BA (Honors) degree in mathematics from the University of Madras. In 1943, he completed his MA degree from Presidency College. The same year, he joined Madras Presidency College as a part-time lecturer and also started doing research under the supervision of the famous mathematician K. Ananda Rau of Madras University. In 1946, he completed his research work and obtained the Ph.D. degree from the University of Madras. His initial research work involved summability method, in particular Bessel summability method. From the very beginning, K. Chandrasekharan's mathematical interests centered around mathematical analysis and analytic number theory.

At that time in Madras, the most well-known and influential mathematicians in the city were K. Ananda Rau and R. Vaidyanathaswamy of the newly created mathematics department of Madras University and Reverend Father Racine of the Loyola College. Ananda Rau, as has already been mentioned, was a contemporary of the iconic Srinivasa Ramanujan and was trained by G. H. Hardy at the University of Cambridge. K. Ananda Rau initiated K. Chandrasekharan to analytic number theory. Father Racine had obtained his doctorate working under the guidance of E'lie Joseph Cartan (1869–1951) in Paris. He and Vaidyanathaswamy exposed Chandrasekharan to other branches of mathematics. R. Vaidyanathaswamy had interest in many branches of mathematics. He was the first mathematician in India to write a book on topology (published by Chelsea). Chandrasekharan had great respect for Vaidyanathaswamy and in later years, he mentioned that he had studied mathematical logic under the supervision of Prof. Vaidyanathaswamy.

Shortly after completing his Ph.D. degree, K. Chandrasekharan came in contact with a leading American mathematician named Marshall Stone (1903–1989) who was on a brief visit to Madras. Stone was much impressed by Chandrasekharan's remarkable mathematical aptitude and prowess. He took personal initiative and made arrangements so that Chandrasekharan could go to the Institute for Advanced Study in Princeton, USA. He also made plans so that Chandrasekharan got the opportunity to work as an assistant to the mathematician and physicist Hermann Weyl (1885–1955). During his stay at Princeton, Chandrasekharan became very well acquainted with John Von Neumann (1903–1957). He also extensively collaborated with Salomon Bochner (1899–1982) of Princeton University. They worked mainly on analysis centering around Fourier transforms. They collaboratively wrote and published a book *Fourier Transforms*.[135] Just to give an idea about the type of work Chandrasekharan did in Fourier series, it may be relevant to quote his own remark. In his paper titled "On the summation of multiple Fourier series I",[136] he wrote:

> While a single Fourier series and, to a lesser extent, double Fourier series has been investigated in great detail, multiple Fourier series has not received the same degree of attention.

.

---

[135]Princeton University Press, 1950.
[136]*Proceedings of the London Mathematical Society*, 50, (1948), 210–222.

In his work on the summation of multiple Fourier series by spherical means, which was initiated by S. Bochner, it may be noted that the definition of summation by spherical means is definitely more general than that of S. Bochner.

In 1945, the Tata Institute of Fundamental Research (TIFR) was founded in Bombay by Homi J. Bhabha. In the initial years, D. D. Kosambi and F. W. Levi were the faculty members in mathematics. Levi had left Germany during the oppressive reign of the Nazi regime and had been a professor in the prestigious Department of Pure mathematics in the University of Calcutta. After the end of the Second World War, Levi stayed for a short while at the TIFR and left for Germany to rejoin his earlier academic position. Kosambi, on the other hand, was a good mathematician and a versatile scholar. He had great command and interest in disciplines as diverse as differential geometry, statistics, Sanskrit, and Marxist history. But he lacked the leadership quality for building up a new academic center.

Shortly after establishing the TIFR, Bhabha was on the lookout for recruiting very good scholars in different disciplines for the Institute. During his visit to the Institute for Advanced Study in Princeton, he met Chandrasekharan there. On the dual recommendations of Hermann Weyl and Von Neumann, Homi Bhabha made up his mind and invited K. Chandrasekharan to join the TIFR. K. Chandrasekharan accepted the offer. In 1949, he joined the Institute as a reader in mathematics. Soon after taking up the responsibilities at the TIFR, he invited S. Minakshisundaram (1913–1968), a leading Indian mathematician of the twentieth century to join the Institute. Both of them collaborated and worked in mathematical analysis which led to the publication of a book titled *Typical Means*.[137]

Chandrasekharan was a great mathematician but his contributions in building up the world renowned school of mathematics at TIFR is amazing. He systematically organized the different mathematical activities there. The culture to which he had been exposed at the Advanced Center at Princeton influenced him and he tried to emulate that pattern. He planned to build something in the line of a graduate school of an American university where students initially study some basic subjects and then move on to specialize on a particular discipline for acquiring a Ph.D. degree. At that time, another famous Indian number theorist named K. G. Ramanathan earned his Ph.D. degree from Princeton University working under the supervision of Carl Ludwig Siegel (1896–1981) at the Institute for Advanced Study. K. Chandrasekharan invited K. G. Ramanathan and Ramanathan accepted the offer and joined the TIFR. With the help of Ramanathan, Chandrasekharan started the graduate programme at TIFR. From the whole of India, on the basis of interviews, a few talented students were selected each year. In order to facilitate the training programme and train up students properly, Chandrasekharan undertook a novel programme of getting a large number of visiting professors from abroad. With his earlier contacts at Princeton, he successfully persuaded many leading mathematicians of the twentieth century from Europe and the USA to visit TIFR for long periods of time and teach the graduate courses.

---

[137]Oxford University Press, (1952).

This programme of inviting foreign scholars started from the year 1953. Some of the famous mathematicians who visited the Institute during the initial years include C. L. Siegel (1896–1981), S. Eilenberg (1913–1998), H. Rademacher (1892–1969), M. Eichler (1912–1992), O. Zariski (1899–1986), and L. Schwartz (1915–2002). The lecture notes of such eminent mathematicians delivered during specialized courses were taken down by the students attending the courses. Later, these written up notes were published as *TIFR Lecture Notes Series*. Several talented students joined TIFR and for a number of years the faculty of the TIFR consisted of these students who made a mark in their respective fields of specialization. TIFR gradually acquired the reputation of being a world-class center for mathematics. Another important initiative of K. Chandrasekharan was to organize every four years an international colloquium on various topics in mathematics. These were closed-door meetings of invited experts from all over the world on a specific topic, meant for all TIFR faculty and students. The topic and invitation to relevant experts were chosen by the members of the mathematics faculty. Chandrasekharan persuaded the International Mathematical Union (IMU) to sponsor such colloquiums and involved the Dorabji Tata Trust to take care of the required expenses.

The first such colloquium was held in 1956 on "*zeta functions*". Atle Selberg (1917–2007), the famous number theorist, delivered four talks in this colloquium. The "Selberg trace formula" [22] was discussed in these lectures.

Incidentally, it is worth mentioning that around this time, K. Chandrasekharan became interested on the zeta function. H. Hamburger (1889–1956) had proved a certain result. By connecting the functional equation to the modular relation, C. L. Siegel had obtained a simple proof. Bochner and Chandrasekharan in their publication titled "On Riemann's functional equation"[138] raised some questions and for answering those they associated a constant δ_λ with any sequence $\lambda = (\lambda_n)$ and were successful in obtaining an upper bound for the number of solutions in terms of d_λ and d_μ. For a particular case, they proved that $\delta = 1$ or 3 and tabulated all the corresponding solutions. In another publication, with S. Mandelbrojt, K. Chandrasekharan[139] discussed in detail the important cases of $\delta = 1$ and $\delta = 3$.

Professor Chandrasekharan supervised Raghavan Narasimhan (1926–2007), who later became a famous number theorist. Chandrasekharan in collaboration with Narasimhan did a lot of work on analytic number theory. Starting from 1960 in 18 years (up to 1978), they contributed as many as 13 research papers which were published in famous international journals and reviewed suitably.

In the first four papers, they worked on Hecke's functional equation and also discussed the average order of arithmetical functions, arithmetical identities with relation to that. In a 1962 publication titled "Functional equations with multiple gamma factors and the average order of arithmetical functions",[140] the two authors wrote an asymptotic formula for the Riesz means of the coefficient as the main term $M(x)$ plus an error term $E(x)$. The term $M(x)$ is given by an integral. They successfully

---

[138] Ann. Math., 63, (1956), 336–360.

[139] Ann. Math., 66, (1957), 285–296.

[140] Section 3, (KC. 5).

determined the O-result and the Ω-result in this generality. Then these results were applied by them to the classical arithmetic functions like $d(n)$, $\sigma(n)$, and $\varphi(n)$. In a paper published in 1964,[141] Chandrasekharan and Narasimhan continued their investigations and obtained L_2-estimate for the error term. Later other researchers have extensively used this result.

It may be noted that G. H. Hardy and J. E. Littlewood had much earlier deduced an approximate functional equation for the Riemann zeta function. This result was used by Chandrasekharan and Narasimhan to obtain an asymptotic formula for L_2-mean and an upper bound for L_4 mean for the Riemann zeta function on the critical line. Realizing the importance of their work and inspired by its usefulness, Chandrasekharan and Narasimhan published a paper in 1963[142] and deduced an approximate functional equation for any Dirichlet series satisfying a functional equation with multiple gamma factors under some mild conditions. They applied these results to the Dedekind zeta functions and deduced L_2-estimate for the same on the critical line. In the study of Zeta functions occurring in algebraic number theory, this result has been very useful.

The works of K. Chandrasekharan and R. Narasimhan on O- and Ω-results for Riesz means, mean values of error terms for the summatory functions of a wide class of arithmetical functions, and on approximate functional equations are very important contributions in the field of analytic number theory.

Appendix

[22] *Selberg trace formula*: First introduced by A. Selberg in 1956, this formula eponymously named after him gives an expression for the character of the unitary representation of G on the space $L^2(G/\Gamma)$ of square-integrable functions, where G is a Lie group and Γ is a cofinite discrete group. The character is given by the trace of certain functions on G.

2.4.2 Kollagunta Gopalaiyer Ramanathan (K. G. Ramanathan) (1920–1992)

Professor K. G. Ramanathan (Fig. 2.7) was a very distinguished Indian number theorist who along with Prof. K. Chandrasekharan was a key figure in initiating the school of mathematics at the Tata Institute of Fundamental Research (TIFR), and also for developing one of the most outstanding schools of number theory there.

K. G. Ramanathan was born on November 13, 1920, in the city of Hyderabad (now, Telangana). His father was K. Gopala Iyer and mother Smt. Anantalakshmi. He had his school education from Wesleyan Mission High School of Secunderabad. In 1940, he graduated with a B.A. degree in mathematics from the Nizam College under the Osmania University of Hyderabad. Thereafter, he went to Madras (now, Chennai)

---

[141] Section 3, (KC. 8).

[142] Section 3, (KC. 7).

Fig. 2.7 K.G. Ramanathan
(1920–1992)

for carrying on further higher studies in mathematics. In 1942, he completed his
M.A. in mathematics from Loyola College under the University of Madras. It was
at that college that he was taught by Reverend Father Racine. After completing his
postgraduation, during 1945–1946, he worked as an assistant lecturer in mathematics
at the Annamalai University at Chidambaram. For the next two years, he worked as a
Lecturer at Osmania University, Hyderabad. Then in 1948, he returned to Madras. He
joined as a research scholar at the Department of Mathematics at Madras University.
During that time, he came in contact with well-known renowned mathematicians
Prof. R. Viadyanathaswamy and Dr. T. Vijayaraghavan. The Reverend Father Racine
of Madras Loyola College, whom he had known earlier, inspired Ramanathan to
pursue research in higher mathematics. While doing his postgraduate studies, he
started doing research on number theory. Even before completing his M.A. examination, in 1941, he published his first paper on number theory. In this research paper
titled "On Demlo numbers",[143] he looked into the problem of describing the digits
of the product of two factors in terms of the digits of the factors. He also examined
some special cases. Some other papers that he published during these four years
(1941–1945) were as follows:

- "Congruence properties of $\sigma(n)$, the sum of the divisors of n" [Sect. 2.3, (KGR.
  2)]
- "On Ramanujan's trigonometrical sum $C_m(n)$" [Sect. 2.3, (KGR. 4)]
- "Some applications of Ramanujan's trigonometrical sum $C_m(n)$" [Sect. 2.3,
  (KGR. 6)]
- "Multiplicative arithmetic functions" [Sect. 2.3, (KGR. 3)]
- "Congruence properties of Ramanujan's function $\tau(n)$" [Sect. 2.3, (KGR. 5)]
- "Congruence properties of σ_a (n)" [Sect. 2.3, (KGR. 7)]
- "Congruence properties of Ramanujan's function $\tau(n)$ *II*" [Sect. 2.3, (KGR. 8)].

---

[143] Section 3, (KGR. 1).

All these papers were reviewed by the famous number theorist D. H. Lehmer. Reviewing the paper published in 1945 on Ramanujan's function $\tau(n)$, Lehmar wrote:

> This paper is concerned with a sum which is, in fact, the sum of the n-th powers of the primitive mth roots of unity. The author points out its connection with partitions m, the sum in question being the excess of the number of partitions of n into an even number of incongruent parts modulo m over those into an odd number of such parts. Simple proofs are given of a number of known theorems, such as the one that asserts that the product of 2 sin $\pi(n/m)$ taken over all n less than and prime to m has the value p or 1 accordingly as m is or is not a power of the prime p.

Shortly after that in the mid-forties of the twentieth century, K. G. Ramanathan went to the USA. There he joined the Institute for Advanced Study in Princeton and worked as an assistant to Prof. Hermann Weyl. He subsequently came in contact with the internationally famous number theorist Prof. C. L. Siegel and was greatly influenced by him. This was the turning point of his career. Siegel's influence was long-standing and was noticed in his subsequent research activities. At Princeton University, Ramanathan worked under the supervision of the famous number theorist Prof. Emil Artin (1898–1962) and obtained his Ph.D. degree in 1951 for his doctoral thesis titled "The Theory of Units of Quadratic and Hermitian Forms".

Soon after that, Ramanathan came back to India and in 1951 he joined the TIFR in Bombay. It has already been discussed earlier how he joined hands with Prof. K. Chandrasekharan and built up the School of mathematics at TIFR. But Ramanathan's role in setting up an excellent school of research on number theory deserves special mention. He had remarkable expertise and a great passion for number theory. This certainly helped him to achieve his goal. Professor K. G. Ramanathan was a great teacher. He took special care to prepare his lectures and in the class, he delivered them very lucidly. The informal discussions that he encouraged in the class unfolded for many an aspiring scholar grand vistas of the exciting mathematical world of the greatest number theorists of modern times such as Fermat, Euler, Lagrange, Gauss, Abel, Jacobi, Dirichlet, Kummer, Galois, Eisenstein, Kronecker, Riemann, Dedekind, Minkowski, Siegel, Hilbert, Hecke, Artin, Weil, and so on. His abiding enthusiasm and an endless passion to propagate good mathematics as well as refined mathematical culture benefitted not only TIFR but was also responsible for the general improvement of mathematical teaching and research in India.

K. G. Ramanathan was a highly acclaimed Indian number theorist of the twentieth century. His contributions received international recognition. His best research works are related to the analytic and arithmetic theory of quadratic forms over involutorial division algebras. At the beginning of his career, during 1943–1950, Ramanathan worked mainly with congruence properties of some arithmetic functions, Ramanujan's trigonometrical sums, and certain identities of the Ramanujan type. During 1951–1952, he worked on quadratic forms. The influence of Siegel's work on quadratic forms together with P. Humbert's (1891–1953) reduction theory led him to publish the following two papers. They are

- "The theory of units of quadratic and Hermitian forms" [Sect. 2.3, (KGR. 12)]
- "Units of quadratic forms" [Sect. 2.3, (KGR. 14)].

In the papers mentioned above, he studied the properties of unit groups of quadratic and Hermitian forms over algebraic number fields, such as their finite generation or finiteness of their convolute. His abiding interest in quadratic forms resulted in further publications on related topics in 1956. In another important research work published in 1959,[144] Ramanathan has used a general formula of Siegel concerning lattice points in specified domains and obtained a formula for the discriminant of the division algebra. This led to the well-known Hasse–Brauer local–global splitting theorem [23] for quaternion algebras [24] over the rationals.

His 1961 publications of two papers are important. In the first paper,[145] Ramanathan made a systematic study of the equivalence of representation by quadratic forms over division algebras [25] with involution. In the second paper of the series,[146] he considered the case of the theta series associated with the abovementioned quadratic forms.

Using ideas of his own earlier results along with certain techniques of Siegel and some theorems of A. Selberg and A. Borel, Ramanathan did some notable work on discontinuous groups. He made two research publications on these investigations. In the first of these two papers,[147] Ramanathan solved the problem of combining infinitely many classes of mutually incommensurable discrete groups of the first kind in classical semisimple groups. In the second paper,[148] he settled the question regarding the maximality of discrete subgroups of arithmetically defined classical groups, thus generating certain results of Hecke and Maass.

In collaboration with his student and noted number theorist S. Raghavan, Ramanathan published three papers dealing with Diophantine inequalities and quadratic forms. Jointly with S. Raghavan, he proved an analogue over algebraic number fields of a result of A. Oppenheim on the density of values of indefinite quadratic forms, in $n \geq 5$ variables, that are not scalar multiples of rational forms and represent zero (which means they admit non-trivial integral solutions).

From 1974 onwards till 1990, K. G. Ramanathan became actively involved in the published and unpublished works of Srinivasa Ramanujan. His main areas of interest were Ramanujan's beautiful work on singular values of certain modular fractions, Rogers–Ramanujan continued fractions, and hypergeometric series. He tirelessly tried to urge many of his colleagues in India to seriously take up studies on Ramanujan's fascinating unpublished works. In spite of his failing health and resulting disability, he continued to work on a "monograph on Ramanujan's continued fractions" highlighting their twin aspects related to hypergeometric series and basic hypergeometric series, respectively.

Professor K. G. Ramanathan's personal research contributions were more than 45. He was also the author of two books on advanced mathematics, one of which was written singly by him and the other in collaboration with other mathematicians.

---

[144] Section 3, (KGR. 20).

[145] Section 3, (KGR. 21).

[146] Section 3, (KGR. 22).

[147] Section 3, (KGR. 23).

[148] Section 3, (KGR. 24).

As already mentioned and discussed earlier, he was a number theorist of the highest order and was one of the architects of the number theory school of TIFR. A quotation from the "K. G. Ramanathan Memorial Issue" of the *Proceedings of the Indian Academy of Sciences*, published in February 1994, truly reflects his academic standing:

> Professor K. G. Ramanathan was of small build but had a big influence on the post-independence Indian mathematical scene. Despite the legacy of the legendary Srinivasa Ramanujan and several mathematicians of high standing early in this century, pursuit of mathematics had remained rather weak in India till the fifties. He was one of the few people responsible for the fortification which has put India firmly back on the international mathematical map. He not only was himself a front-ranking mathematician of international reputation, but also contributed a great deal to the emergence of a strong mathematical base at the Tata Institute of Fundamental Research as also to the overall development of research and teaching of mathematics in India, and to an extent, even beyond our shores. He was well recognized for his achievements in number theory, especially the analytic and arithmetic theory of quadratic forms over division algebras with involution.

Appendix

[23] *Brauer–Hasse theorem*: The Albert–Brauer–Hasse–Noether theorem establishes a local–global principle for the splitting of a central simple algebra A over an algebraic number field K.

[24] *Quaternion algebra*: A quaternion algebra over a field F is a central simple algebra A over F that has dimension 4 over F. By extending scalars (similarly, tensoring with a field extension), every quaternion algebra becomes the matrix algebra.

[25] *Division algebra*: A *division algebra* which is also called a *division ring* or a *skew field* is a ring in which every non-zero element has a multiplicative inverse, but multiplication is not necessarily commutative.

2.4.3   Srinivasacharya Raghavan (1934–2014)

Srinivasacharya Raghavan (or S. Raghavan) was born on April 11, 1934, in Thanjavur district (now, Tamil Nadu). After completing his school education from the town of Palayamkottai, in 1954, he obtained his B.A. (Honors) degree in mathematics from St. Joseph's College at Trichinapally (now, Tiruchirapally). Soon after that, he was selected and joined the famous School of mathematics at the Tata Institute of Fundamental Research (TIFR) in Mumbai as a research student (Fig. 2.8). He specialized in number theory under the joint supervision of Profs. K. Chandrasekharan and K. G. Ramanathan.

Professor Raghavan started publishing research papers on various topics of number theory from 1959 onwards. In 1960, he obtained his Ph.D. degree from the University of Bombay. Thereafter, he became a faculty member of TIFR and had a long and illustrious career at the Institute. His outstanding research performance

Fig. 2.8 S. Raghavan

helped him to become a professor at a relatively young age of 41 in the year 1975. He finally took voluntary retirement as a senior professor of mathematics in 1994.

In the early days of his research career in the late sixties of the twentieth century, S. Raghavan collaborated with S. S. Rangachari of TIFR. The central topic of their research was quadratic and modular forms. They made detailed investigations on Ramanujan's integral identities. In collaboration with Rangachari, R. Narasimhan, and Sunder Lal, S. Raghavan was involved in writing mathematical pamphlets on algebraic number theory which was published by TIFR, Mumbai, in 1966. During 1969–1970, Raghavan and Rangachari published two research papers related to quadratic and modular forms. These were published in the *Journal of the Indian Mathematical Society* and *Acta Arithmetica*, respectively. In the eighties of the twentieth century (1980–1981), Raghavan and Rangachari jointly published two more papers on the Poisson formula of Hecke type. In 1989, they jointly published a paper related to Ramanujan's elliptic integrals and modular identities. This write-up on Ramanujan's elliptic integrals and modular identities was published by TIFR in 1988–1989. This work is of special importance. Raghavan did a lot of research on Ramanujan's identities. Using the theory of modular forms, he gave proofs for some of the identities. In 1997, he delivered a series of lectures on Ramanujan's work at Madurai Kamaraj University in Tamil Nadu. The contents of his lectures were written as a technical report by him. He also wrote an article highlighting the impact of Ramanujan's work on the post-Ramanujan and modern-day mathematics. Like many Indian mathematicians of the twentieth century, S. Raghavan was highly motivated by Srinivasa Ramanujan.

Going back to his other research activities, it may be noted that under the joint guidance of Chandrasekharan and Ramanathan, he soon learnt and mastered Siegel's theory of modular forms of degree 3. He applied it effectively to generalize the classical problem of representation of positive integers as values of positive quadratic forms. In the process, his work involved the generalization of the earlier results of Hardy, Ramanujan, Hecke, and Peterson. His first research paper[149] contains this important work. It dealt with the representation of modular forms of degree n by quadratic forms.

Probably influenced by K. G. Ramanathan, in subsequent years in his long academic career spread over more than 50 years, Raghavan's main interest of research was centered around modular forms, automorphic functions [26], quadratic forms, and Hermitian forms. He investigated them from various angles and pointed out their interrelations. In his work, one regularly comes across such themes as an estimation of Fourier coefficients of Siegel modular forms, a variety of structural aspects of modular forms, and the distribution of values of quadratic forms. According to the renowned mathematician Prof. S. G. Dani:

> His (Raghavan's) application of Hecke's Grenzprozess to analytic continuation of non-holomorphic Eisenstein series of degree 3 became a forerunner of Weissauer's deep generalisation for general n.

Among his other important research initiatives, Raghavan took special interest in Oppenheim conjecture [27] on the density of values of indefinite quadratic forms, which happens to be an important conjecture in number theory. In collaboration with K. G. Ramanathan, he successfully established an analogue to the said conjecture over algebraic number fields. His work on Oppenheim conjecture for forms that might not represent zero indirectly influenced young research workers at the TIFR of those times. M. S. Raghunathan and S. G. Dani did some remarkable work in this area. In turn, their work was to influence the pathbreaking research carried out by G. A. Margulis and Marina Ratner on related topics. Later, S. Raghavan collaborated with S. G. Dani and conducted noteworthy research on the density of orbits of irrational Euclidean frames under the action of different types of familiar discrete groups of significance in Diophantine approximation of systems of linear forms.

Finally, in the context of his career, it may be noted that throughout his academic and professional life, he was a superlative performer. Ever since he entered the TIFR as a student, he became an integral part of the mathematical research community of the Institute. During his long and illustrious career, S. Raghavan was visiting scientists to various prestigious academic centers of mathematical excellence. As an academician, he visited the Institute for Advanced Study, Princeton, USA, during 1986–1987. At home, he was a visiting professor at the University of Bombay for 6 years from 1968 to 1974. During 1991–1994, he was a visiting professor at SPIC Mathematical Institute, Chennai. This Institute is presently known as the Chennai Mathematical Institute.

Apart from academic assignments, S. Raghavan made important collaborative research with R. J. Cook of the University of Sheffield in England. They worked on the values of quadratic forms. In collaboration with J. Sengupta of TIFR, he

---

[149]Section 3, (SR. 1).

wrote several research papers and published them. These papers dealt with modular forms, specifically on Fourier coefficients of Maass cusp forms in three-dimensional hyperbolic space. To his own credit, Prof. S. Raghavan had published about 50 research papers related to various topics on number theory. Four students had obtained their Ph.D. under his guidance.

In 2014 he breathed his last in Chennai.

Appendix

[26] *Automorphic functions:* These are functions on a space that are invariant under the action of some group. Alternately, it may be defined as a function of the quotient space. Often, the space is a complex manifold and the group is a discrete group.

[27] *Oppenheim conjecture*: In Diophantine approximation "Oppenheim conjecture" concerns representations of numbers by real quadratic forms in several variables.

2.4.4   Kanakanahalli Ramachandra (1933–2011)

Kanakanahalli Ramachandra (or K. Ramachandra) was born on August 18, 1933, in Mandya in the state of Mysore (now, Karnataka) in the South of India (Fig. 2.9). He came from a family of modest means and lost his father at the tender age of 13. His mother mortgaged their small agricultural property and with the loaned money managed to take care of her son's education.

As a student, Ramachandra had been awarded a brief biography of the iconic mathematical genius Srinivasa Ramanujan, as a prize in a competition. That book acted as a catalyst and ignited the interest for mathematics in the mind of young Ramachandra. A small anecdote about student Ramachandra would make for interesting reading and also demonstrate his early promise as a great mathematician of the future. To mathematicians with some knowledge of Ramanujan, his taxi-cab number $1729 = 9^3 + 10^3 = 1^3 + 12^3$ is very well-known. As a college student, Ramachandra was involved in a similar episode. The Principal of his college had a car with the number 3430 on its number plate. Ramachandra painstakingly investigated the mathematical possibilities of the number and he eventually discovered that upon adding 5, the number 3435 happens to be the only number with the unique property that when each digit was raised to a power equal to itself and the resulting numbers were added up, the sum equals the original number. That is $3^3 + 4^4 + 3^3 + 5^5 = 3435$. For a young college student, this was indeed a remarkable discovery. Ramachandra had completed both his graduation and postgraduation from Central College, Bangalore (now, Bengaluru). During his student days at Central College, in a public library in the town, he came across a copy of G. H. Hardy's famous lecture series on Ramanujan's research works. It was published as a book titled *Ramanujan: Twelve Lectures on subjects suggested by his Life and Work*. Ramachandra studied the entire book with great interest and concentration. This particular book certainly

(a)

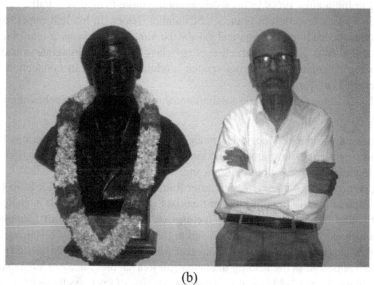

(b)

Fig. 2.9 Kanakanahalli Ramachandra (1933–2011): **a** taking a class, **b** with a statue of Srinivasa Ramanujan

inspired Ramachandra to take up a research career in mathematics and especially on number theory. Till the end of his life, Ramachandra preserved a copy of this book, and it was one of his personal favorites.

Through an initial period of temporary low-grade jobs, he worked as a lecturer for some time in an engineering college and served as a teacher for a week in the Indian Institute of Science. After struggling for a couple of years, in 1958, he was finally selected for graduate studies at the TIFR. There he met Prof. K. Chandrasekharan who was at that time recognized as one of the experts in the theory of the Riemann zeta function in India. Ramachandra studied this theory with great care and devotion under the guidance of Prof. Chandrasekharan. In future, Ramachandra himself became one of the leading experts in the field of Riemann zeta function. He made notable contributions in this branch of number theory.

Ramachandra worked as a research professor at TIFR for nearly three long decades until his retirement in 1995. In a sense, he was perhaps the true successor of Srinivasa Ramanujan in modern Indian mathematics. Taking various incidents into account, it would not be an exaggeration to say that without the relentless pursuits and efforts of Prof. K. Ramachandra, number theory would have been extinct in India in the middle of the seventh decade of the twentieth century.

Professor Ramachandra is arguably the best analytic number theorist to have worked in India after 1960. In his work, spanning more than almost half a century, he made a mark in different branches of number theory. In his first paper, which appeared in the leading mathematical journal the *Annals of mathematics,* he obtained remarkable results on the construction of ray class fields of an imaginary quadratic field, using calculations involving special modular functions in connection with Kronecker's limit formula.

Professor Ramachandra made notable contributions in various branches of number theory.

1. *Elementary number theory*: He made serious investigations on various questions involving summatory functions of several arithmetical functions, on Vinogradov's three primes theorem, etc.
2. *Algebraic number theory:* A year after Ramachandra joined the TIFR, in 1959, the internationally famous number theorist C. L. Siegel (1896–1987) visited the TIFR and delivered a course of lectures on the Kronecker limit formula. K. Ramachandra attended the course and at the suggestion of K. G. Ramanathan, looked into the applicability of Kronecker's limit formulae [28]. In his first paper, which appeared in the leading mathematical journal the *Annals of mathematics,* he obtained remarkable results on the construction of ray class fields of an imaginary quadratic field, using calculations involving special modular functions in connection with Kronecker's limit formula.

This paper titled "Some applications of Kronecker's limit formula"[150] was highly praised by the reviewer M. Eichler. In his comments, the reviewer wrote:

---

[150]Section 3, (KR. 1).

This paper contains some remarkable new results on the construction of the ray class field of an imaginary quadratic number field.

Ramachandra made a remarkable application of Kronecker's second limit formula to the theory of complex multiplication. He used it for the construction of a certain maximally independent set of units in a given class field of an imaginary quadratic field and the evaluation of a certain elliptic integral. This elliptic integral was originally found by Chowla and Selberg and also by Ramanujan.

The calculation of units of the cyclotomic field generated by the pth root of unity for a prime number p is classical computation as carried out by E. Kummer. The question of studying the units in the cyclotomic field generated by mth root of unity is a more difficult problem. Ramachandra took up the case and the units constructed by him have come to be known as Siegel–Ramachandra–Robert units. For Abelian extensions of quadratic imaginary fields constructed by using elliptic functions, the Siegel–Ramachandra–Robert units are explicit units. Siegel did the initial work and they were constructed by Ramachandra in the sixties. Later, they were streamlined by Robert. Perhaps that is why they were called Siegel–Ramachandra–Robert units for some time. These elliptic units have a very important role in many investigations dealing with arithmetic and elliptic curves with complex multiplication. The fundamental works of Coates–Wiles and Rubin deserve special mention in this context.

It is believed that for every integer n that is large enough, the interval of length $(n)^{1/2}$ starting from n must contain a prime number, but the existing state of knowledge at that time was not adequate to prove the result, even if the Riemann hypothesis was used. Ramachandra posed the problem to study if the interval always contained an integer with a "large" prime factor. He proved that it was in fact so and that there was an integer m in the interval having a prime factor that exceeded $m^{(1/2)+(1/3)}$. Ramachandra's method involves clever use of the sieve methods and also exponential sum estimates, originally due to Van der Corput. Ramachandra's problem has attracted the attention of a number of leading number theorists in the subject and remains an important question even today.

Incidentally, it may be relevant to mention here that in the work on elliptic curves and curves of higher genus, many Indian researchers have contributed in a significant way. On the explicit determination of the number of points on an elliptic curve with complex multiplication, the number theorists from the Panjab school from Chandigarh comprising A. R. Rajwade, M. K. Agarwal, J. C. Parnami, D. B. Rishi, S. A. Katre, and the students of Prof. R. Balasubramanian namely R. Padma and S. Venkataraman from the MatScience (new name, Institute of Mathematical Sciences), Madras (Chennai) have contributed a large number of research papers.

3. *Transcendental number theory*[29]: In the sixth decade of the twentieth century, Alan Baker published a seminal work on transcendental functions [30]. He made important advances in transcendental number theory that earned him the Fields Medal. After this, K. Ramachandra also started to take great interest in this branch of number theory. His first result in this theory, proved about the same time by

Prof. Serge Lang, is called "Six Exponentials Theorem" [31]. In transcendental number theory, this particular theorem is a result that, given the right conditions on the exponents, guarantees the transcendence of at least one set of exponentials. This remains a fundamental result in this theory. Sometime in the 1960s, the theorem was explicitly stated and completely proved independently by S. Lang and K. Ramachandra. The detailed discussions on Ramachandra's contributions can be found in Michel Waldschmidt's paper titled "On Ramachandra's contributions to Transcendental Number Theory".[151]

Ramachandra's other research contributions in this area are based on Baker's theory on linear forms and logarithms, transcendental measures of certain irrational numbers, and so on. Ramachandra and his students, especially T. N. Shorey, seriously pursued research on transcendental number theory and made remarkable contributions to both the theory and its applications to problems of classical number theory.

Baker's main result may be recalled. If $a_1, a_2, ..., a_n$ are logarithms of algebraic numbers, then Baker showed that if the a_i are linearly independent over the field of rational numbers, then they are linearly independent over the field of algebraic numbers. Baker, in fact, got a lower bound for the linear combination of the $a_1, a_2, ..., a_n$. This lower bound of Baker was improved by many mathematicians, including K. Ramachandra.

Ramachandra is better known for applications of Baker's theory to arithmetical questions. An attempt is made to explain one such question. Let $n + 1, n + 2, n + 3, ..., n + k$ be a set of consecutive integers. One would like to know if there exist distinct primes $p_1, p_2, ..., p_k$ such that the prime p_j divides $n + j$ for every j. C. A. Grimm conjectured in 1969 that if the set of the consecutive numbers does not contain a prime number, then this would be always possible. Using Baker's theory of linear forms, Ramachandra, Shorey, and Tijdeman proved that if k is at most $c(\log n)^3/(\log \log n)^3$, then Grimm's conjecture holds. It may be noted that this result established in 1974 has resisted all attempts at improvement to this day.

4. *Theory of Riemann zeta function and L-functions*: After publishing some important papers on the Hardy–Ramanujan circle method and a paper related to Thue–Siegel–Roth theorem, from 1973 onwards K. Ramachandra turned his attention to hardcore classical analytic number theory, especially the theory of Riemann zeta function and general Dirichlet series. He has vast contributions in this area. To name a few, he has established Omega theorems, lower and upper bound estimates on various questions, zero-density estimates of certain L-functions, mean-value theorems on certain vertical lines of certain L-functions, and so on.

Some of his remarkable results include a simple proof of an asymptotic formula for the fourth power mean of Zeta function. This method is now widely used for getting mean square estimates for the Dirichlet series. He also showed by a method, which

---

[151]Ramanujan Mathematical Society, Lecture Notes Series Number 2, (2006).

he called Hooley–Huxley contour, to use Perron's formula even for those Dirichlet series which may have singularities.

The other important contribution (some of which was done in collaboration with his student R. Balasubramanian and M. V. Subbarao) is to get a lower bound for the L^1, L^2 mean of the Dirichlet series. He used this to obtain Omega results for the Riemann zeta function on the line $s = \frac{1}{2}$. He also successively used this concept to get an Omega result for the error term in the summatory functions of arithmetic functions.

Ramachandra's contributions to the theory of Riemann zeta function have been nicely described by the British mathematician Roger Heath-Brown who is a Fellow of the Royal Society. Addressing Prof. Ramachandra, he wrote:

As soon as I entered research, 30 years ago, yours became a familiar name; and your influence has remained with me ever since. Time permits me to mention in detail only one strand of your work—but it is one that clearly demonstrates how important your research has been. A little over 20 years back you proved the first results on fractional moments of the Riemann–Zeta function. At first, I could not believe they were correct! Since then however the ideas have been extended in a number of ways. They have led of course to a range of important new results about the Zeta function and other Dirichlet series. But just as significantly the ideas have led to new conjectures on the moments of the Riemann–Zeta function. These conjectures provide the first successful test for the application of random matrix theory in this area. Nowadays this is a growing area which has contributed much to our understanding of Zeta functions. And it can all be traced back to your work in the late 1970s.

Professor Ramachandra was a great teacher and he successfully mentored some of the brightest number theorists of India. R. Balasubramanian and T. N. Shorey are two of his most successful students who have been regarded as outstanding number theorists for their remarkable contributions in their respective fields. T. N. Shorey, S. Srinivasan, R. Balasubramanian, M. J. Narlikar, V. V. Rane, A. Sankaranarayanan, K. Srinivas, and Kishore Bhat obtained their respective Ph.D. degrees under the supervision of K. Ramachandra.

Appendix

[28] *Kronecker limit formula*: The classical "Kronecker limit formula" describes the constant term at $s = 1$ of a real analytic Eisenstein series (or Epstein zeta function) in terms of Dedekind eta function.

[29] *Transcendental number theory*: It is a branch of number theory that investigates transcendental numbers (numbers that are not solutions of any polynomial equation with integer coefficients) in both qualitative and quantitative ways.

[30] *Transcendental function*: It is a type of function which is not expressible as a finite combination of the algebraic operations of addition, subtraction, multiplication, division, raising to a power, and extracting a root, for example, $\log x$, $\sin x$, $\cos x$, e^x, or any function containing any of them.

[31] *Six exponentials theorem*: In transcendental number theory, the "six exponentials theorem" is a result that, under given conditions on the exponents, guarantees the transcendence of at least one set of exponentials.

Another very talented number theorist who made remarkable contributions in algebraic number theory in a very short span of life will be discussed now. He is C. P. Ramanujam. Like his namesake, the iconic Srinivasa Ramanujan, he too had a brief tragic life.

2.4.5   *Chakravarti Padmanabhan Ramanujam (1938–1974)*

Chakravarti Padmanabhan Ramanujam (or C. P. Ramanujam), Fig. 2.10, was born on January 9, 1938, in the city of Madras (now, Chennai). His father C. Padmanabhan was an advocate of the Madras High Court. In 1952, he passed his high school examination and joined the famous Loyola College of Madras. In 1957, he graduated with honors in mathematics but failed to get a first class. During his college days, he was taught by Father C. Racine of Loyola College in the honors classes. Ramanujam had a great regard and respect for Father Racine and maintained a regular correspondence with him till the end of his life. On the suggestion of Father Racine, Ramanujam applied to the School of mathematics at TIFR and got selected. Father Racine had given him a good letter of recommendation in which he wrote:

> He (Ramanujam) has certainly originality of mind and the type of curiosity which is likely
> to suggest that he will develop into a good research worker if given sufficient opportunity.

Future developments would vindicate Father Racine's confidence in his student. Before joining TIFR, Ramanujam had a brief exposure to the famous number theorist Prof. T. Vijayaraghavan who was the Director of the Ramanujan Institute of mathematics at Madras till 1954.

Ramanujam's major contributions in number theory are in the area of Waring's problem for number fields. After the successful completion of his initial training at

Fig. 2.10 C.P. Ramanujam
(1938–1974)

the School of mathematics (at TIFR), C. P. Ramanujam started to work under the guidance of Prof. K. G. Ramanathan. He started working on a problem related to Lie groups and differential geometry connected with the work of C. L. Siegel. Early in 1961, he started investigations on a problem regarding Diophantine equations especially related to those over number fields. Earlier, C. L. Siegel had raised a problem regarding the generalization of Waring's problem to algebraic number fields. Davenport and his co-workers D. J. Lewis and B. J. Birch had obtained important results in this context. H. Davenport had established that every cubic form with rational coefficients in at least $g = 32$ variables had a non-trivial rational zero. Using Siegel's generalization of the major and minor arcs of the Hardy–Littlewood–Ramanujan circle method, Ramanujam tried to generalize Davenport's method to cubic forms over algebraic number fields. He first simplified Siegel's method and then was successful in proving that every cubic form in 54 variables over any algebraic number field K had a non-trivial zero over that field. His theorem was later refined by P. Pleasants and C. Hooley.

Ramanujam's theorem is stated as follows.

Theorem 1 (C. P. Ramanujam): Any cubic form over any algebraic number field in ≥ 54 variables has a non-trivial zero.

Refining the method used by C. P. Ramanujam, P. Pleasants showed that Ramanujam's result holds with $g = 16$. This is the strongest theorem established in this area so far. The theorem is stated as follows.

Theorem (P. Pleasants): Any cubic form over any algebraic number field in ≥ 16 variables has a non-trivial zero.

Ramanujam was always interested in Waring's problem in algebraic number fields. A natural question would be if Waring's problem has an affirmative answer in a number field too. But because of local conditions, the problem may not have an affirmative answer. When Ramanujam took up this problem, some interesting results had already been published in that direction. Ramanujam had independently solved the problem raised by Siegel. The theorem due to him is stated as follows.

Theorem 2 (C. P. Ramanujam): Any totally positive algebraic integer in any algebraic number field belonging to the order generated by the mth powers of algebraic integers in that number field is actually a sum of at most $\max(2^m + 1, 8m^5)$ mth powers of totally positive integers.

Later on, C. P. Ramanujam became more interested in algebraic geometry and mathematical analysis. But his contributions in number theory are extremely significant. In 1973, he was elected a Fellow of the Indian Academy of Sciences. Ramanujam was a good teacher and was much loved by his students. But this brilliant mathematician was a victim of a cruel illness that recurred time and again. Finally, he put an end to his life by consuming an overdose of barbiturates and died on October 27, 1974.

Now, some students of Prof. K. Ramachandra, who have made important contributions to the development of research on number theory, are discussed.

2.4.6   S. Srinivasan (1943–2005)

S. Srinivasan was born on September 8, 1943, in Mysore. He had his early education in the town of his birth. He completed his secondary school in 1957 and obtained a B.Sc. degree from the Yuvaraja's College of Mysore. He was a keen student of mathematics all through his student days and pursued a postgraduate course in mathematics from the University of Mysore. He earned his M.Sc. degree in 1964. After spending some years in a small town in South Karnataka, in the late 1960s, he joined the Department of Mathematics at Panjab University at Chandigarh as a research scholar. Finally, in the early seventies of the twentieth century, S. Srinivasan joined the TIFR, Mumbai, to work under the supervision of Prof. K. Ramachandra.

Dr. S. Srinvasan was a reputed number theorist and had a penchant for perfection and simplicity. He worked on various important problems and his methods of solution reflect his understanding of the deep mathematics involved. He attempted to give simple and elegant arguments and managed to get almost the same results. His paper titled "A remark on Goldbach's problem", which was published in the *Journal of Number Theory* in 1980, is considered to be one of his most important contributions in this area. In this particular research paper, he proved certain asymptotic formulae for the Goldbach binary conjecture [32] and the allied twin prime conjecture. To mention a few, his other important works are related to "infinitude of primes" [33], "Brun–Titchmarsh inequality" [34], and "Hilbert's inequality". Simplicity, precision, and perfection are trademarks associated with his research outputs. For a major part of his career, he worked as a faculty member of the Tata Institute of Fundamental Research (TIFR), Mumbai. He retired from there in 2003 and shifted to the TIFR-CAM, Bangalore. There, after many years, he reunited with his mentor Prof. K. Ramachandra. Dr. Srinivasan passed away in 2005.

Appendix

[32] *Goldbach's conjecture*: It is one of the oldest and widely known unsolved problems in number theory. It is also known as *"Goldbach's binary conjecture"* to distinguish it from a weaker conjecture. This yet unproved abovementioned conjecture states that every integer greater than two is the sum of two prime numbers.

[33] *Infinitude of primes*: This is a fundamental statement in number theory which asserts that there are infinitely many prime numbers.

[34] *Brun–Titchmarsh inequality*: In analytic number theory, there is an important theorem called Brun–Titchmarsh theorem. It is named after Viggo Brun and Edward Charles Titchmarsh. This theorem is an upper bound on the distribution of prime numbers in arithmetic progression. The abovementioned inequalities are related to the said theorem.

2.4.7 Tarlok Nath Shorey

Professor Tarlok Nath Shorey (or T. N. Shorey) is a famous student of Prof. K. Ramachandra. T. N. Shorey was born on October 30, 1945. He graduated from DAV College, Amritsar, and in 1965, did his M.A. from Panjab University, Chandigarh. Later in 1968, he joined the TIFR, Mumbai, and started working on number theory under the supervision of Prof. K. Ramachandra. He obtained his Ph.D. in the middle of the seventies of the twentieth century. The topic of his thesis was "Linear Forms in Logarithms of Algebraic Numbers". He joined the TIFR, Mumbai, as a faculty member and finally retired as a senior professor from there.

He is one of India's leading number theorists. His most acclaimed research contributions are on transcendental number theory. He was successful in obtaining the best estimates for linear forms on algebraic numbers. He has also made new applications of Baker's method [35] to Diophantine equations and Ramanujan's τ-functions. Shorey's extensive work on the irreducibility of Laguerre polynomial [36] is significant. In the early nineties of the twentieth century, Shorey and Tijdeman did some important work related to the greatest prime function of an arithmetic progression.

As regards Shorey's contributions to linear form estimates and applications, it may be mentioned that in 1892, J. J. Sylvester (1814–1897) first proved that a product of k consecutive positive integers greater than k is divisible by a prime exceeding k. Using a result of M. Jutila, which depends on estimates for exponential sums and an estimate of linear forms in logarithm, Shorey in a 1974 publication[152] proved that it was sufficient if constant times $k(\log\log k)/\log k$ consecutive integers are taken in place of k consecutive integers in the abovementioned result of Sylvester. This was an improvement on earlier results established by P. Erdös, Q. Tijdeman, K. Ramachandra, and T. N. Shorey. In fact, it is the best known result known to date.

In the sixties, the theory on estimating linear forms in logarithms of algebraic numbers had been developed by A. Baker (1939–2018). The estimation for the linear forms to Baker's theory is an important contribution of Shorey. In a 1974 publication[153] and a 1976 publication,[154] Shorey published a linear form estimate in a more general case. Another very important contribution of Shorey is his linear form estimate related to the conjecture of K. A. Grimm (1926–2018). In his famous conjecture, published in 1969,[155] Grimm had stated that if $x, x + 1, ..., x + k - 1$ are all composite integers,[156] then the number of prime factors of $x(x + 1) \cdots (x + k - 1)$ is at least k. Ramachandra, Shorey, and Tijdeman in their 1976 publication[157]

---

[152]Section 3, (TNS. 5).

[153]Section 3, (TNS. 6).

[154]Section 3, (TNS. 11).

[155]*Amer. Math. Month.* 76 (1969), 1126–1128.

[156]*Composite integer* is a whole number that can be divided exactly by numbers other than 1 and itself. For example, 4 can be divided exactly by 2 as well as 1 and 4. So 4 is a composite number. But 5 cannot be divided exactly except by 1 and itself. So, 5 is not a composite number. It is a prime number.

[157]Section 3, (TNS. 15).

showed that Grimm's conjecture was valid when $(\log x)/(\log x)^2$ exceeds some positive constant. While proving this result, they also showed that the assumption for x, $x + 1, ..., x + k - 1$ to be all composite is not necessary.

Shorey did some good research in the area of applications of linear form estimates to values of polynomials, recurrence sequences, and continued fractions. In the late seventies, he did some significant research in this area. In a 1976 publication,[158] T. N. Shorey in collaboration with R. Tijdeman elaborated on the greatest prime factors of polynomials at integer points. Again in 1977, in an article,[159] Shorey in collaboration with A. J. Vander Poorten, R. Tijdeman, and A. Schinzel applied Gel'fond–Baker method to Diophantine equations. For relatively prime positive integers A and B with $A > B$, it was conjectured that $P(A^n - B^n)/n$ tends to infinity, as n tends to infinity, where $P(k)$ denotes the greatest prime factor of k. In 1904, Birkoff and Vandiver showed that $P(A^n - B^n) > n$ for $n > 6$. In 1962, the result was further improved by Schinzel. In 1975, Stewart confirmed the conjecture for all n with $w(n) \leq \log \log n$, where $w(n)$ is the number of prime divisors of n. In a paper published in 1976,[160] P. Erdös and T. N. Shorey, by applying estimates for linear forms in logarithms, determined lower bounds for $P(A^n - B^n)/n$. In particular, they showed for primes p that $P(2^p - 1) > C_8 p \log p$, where $C_8 > 0$ is an absolute constant. Further, they combined the theory of linear forms in logarithms with "Brun's sieve"[161] to show that $P(2^p - 1) > p (\log p)^2/(\log \log p)^3$ for almost all primes. During 1975–1976, Shorey published two other research papers dealing with linear forms in logarithms and their applications.

Another area where Shorey worked is on irrationality measures and transcendence results. He proved a p-adic[162] analogue of algebraic independence of certain numbers given by an exponential function, a result of Tijdeman.[163] In the early part of the twenty-first century, Shorey in collaboration with Tijdeman did more research in this area. But that is beyond the time frame of this book.

Like many other number theorists of India, T. N. Shorey was also inspired and fascinated by the iconic Srinivasa Ramanujan. He did some significant research work on Ramanujan's τ-function and Ramanujan–Nagell equations. The definition of Ramanujan's τ-function has already been given earlier in this book. In a research publication of 1987,[164] Shorey applied the theory of linear forms in logarithms and proved that $\tau(p^m) \neq \tau(p^n)$, where p is a prime and $\tau(p) \neq 0$; $m > n$ and $m \geq C_{14}$, where C_{14} is an absolute constant. In a paper published in 1987, Shorey in fact established

---

[158] Section 3, (TNS. 13).

[159] Section 3, (TNS. 16).

[160] Section 3, (TNS. 12).

[161] Brun's sieve was developed in 1915 by Viggo Brun. In the theory of numbers, the Brun's sieve is a technique for estimating the size of "sifted sets" of positive integers which satisfy a set of conditions which are expressed by congruences.

[162] It is a special type of number system for any prime number p. Such numbers are considered to be close when their difference is divisible by a high power of p. The higher the power, the closer they are.

[163] *Indag.Math.* 34 (1972), 423–435.

[164] Section 3, (TNS. 35).

an absolute lower bound for the difference of these numbers. Kumar Murty, Ram Murty, and Shorey showed that for non-zero odd integers a, the equation $\tau(n) = a$ has only finitely many solutions in integers $n \geq 1$.[165]

Ramanujan had conjectured that $x^2 + 7 = 2^n$ in integers $x \geq 1, n \geq 1$ had solutions $(x, n) = (1, 3), (3, 4), (5, 5), (11, 7), (181, 15)$. Nagell had proved the same. The equation written above is thus known as the *Ramanujan–Nagell equation*. Shorey and Bugeud collaborated and published a paper dealing with the number of solutions of the generalized Ramanujan–Nagell equation. But the paper was published in 2001 and so it is not considered in the present write-up, because of its twentieth century time frame.

Shorey made notable contributions to problems related to the Nagell–Ljunggren equation. The equation $y^m = (x^n - 1)/(x - 1)$ in integers $x > 1, y > 1, m > 1, n > 2$ is called the Nagell–Ljunggren equation. The equation has solutions given by $(x, y, m, n) = (3, 11, 5, 2), (7, 20, 4, 2), (18. 7, 3, 3)$. It was conjectured that the abovementioned equation has only finitely many solutions. In a paper published in 1986,[166] Shorey showed that the said equation has only finitely many solutions, when n is divisible by a prime congruent to 1 mod m.

Earlier, in 1976 Shorey and Tijdeman[167] had shown that the Nagell–Ljunggren equation had only finitely many solutions, whenever x is fixed. Again in a paper published in 1999,[168] Shorey and his collaborator Saradha showed that the said equation does not hold if $x = Z^2$, such that Z runs through all integers >31 and $Z \in \{2, 3, 4, 8, 9, 16, 27\}$. In the same paper, they also showed that the Nagell–Ljunggren equation implied that x is divisible by a prime congruent to 1 mod m, whenever max (x, y, m, n) exceeds a sufficiently large absolute constant. Later, Y. Bugeud, M. Mignotte, Y. Roy, and T. N. Shorey in a research paper published in 1999[169] showed that the Nagell–Ljunggren equation has no solution with x square.

Another equation which caught the attention of T. N. Shorey was Goormaghtigh's equation. The equation $(y^m - 1)/(y - 1) = (x^n - 1)/(x - 1)$ in integers $x > 1, y > 1, m > 2, n > 2, m > n$ is known as Goormaghtigh's equation. In 1917, Goormaghtigh had shown that

$$31 = \frac{2^5 - 1}{2 - 1} = \frac{5^3 - 1}{5 - 1}; 8191 = \frac{2^{13} - 1}{2 - 1} = \frac{90^3 - 1}{90 - 1}$$

On the basis of these findings, it was conjectured that these are the only solutions to the said equation. Shorey in a paper published in 1989[170] showed that 31 and 8191 are the only primes N with $w(N - 1) \leq 5$, where $w(N - 1)$ is the distinct number of

---

[165] Section 3, (TNS. 38).
[166] Section 3, (TNS. 30).
[167] Section 3, (TNS. 14).
[168] Section 3, (TNS. 81).
[169] Section 3, (TNS. 82).
[170] Section 3, (TNS. 44).

prime divisors of $(N - 1)$, such that all the digits of N are equal to 1 with respect to two distinct bases.

Earlier, Shorey in a paper published in 1986[171] had proved that for positive integers $A, B, x > 1, y > 1$, with $x \neq y$, there are at most 24 integers with all the digits equal to A in their x-adic expansions and all the digits are equal to B in their y-adic expansions.

Six years prior to that in 1980, R. Balasubramanian and T. N. Shorey had published a paper[172] and proved that Goormaghtigh's equation implied that max (x, y, m, n) is bounded by a number depending only on the greatest prime factor of x and y.

Arithmetical progressions and perfect powers is another area of number theory which interested T. N. Shorey. Between 1990 and 1992, T. N. Shorey and R. Tijdeman published three papers: arithmetical progressions, prime factors, and related topics in noted international journals. In this context, a research paper published by Shorey[173] deserves special mention.

Shorey and his collaborator N. Saradha have published quite a few papers on these and related topics in the twenty-first century. That is why they are not discussed in the present write-up. Shorey and his collaborator Saradha and student Shanta Laishram (of Indian Statistical Institute, Delhi) in collaboration with non-Indian researchers have published some papers on different types of extensions of the theorem of Sylvester. But they have all been published in the twenty-first century and so they are not discussed in the current report.

T. N. Shorey and his collaborator N. Saradha did a considerable amount of work in the area of arithmetical progressions with equal products. To start with, it may be mentioned that Erdös and Graham had conjectured that the equation

$$X(X + 1) \cdots (X + K - 1)Y(Y + 1) \cdots (Y + L - 1) = Z^2$$

in integers $K \leq 3, L \leq 3$, and $X \leq Y + L$ has only finitely many solutions in all integral variables $X > 0, Y > 0, Z > 0, K,$ and L. In simplified form, this conjecture implies that

$$x(x + 1) \cdots (x + k - 1) = y(y + 1) \cdots (y + k + 1 - 1)$$

has only finitely many solutions in $x > 0, y > 0, k \geq$, and $1 \geq 0$ satisfying $x \geq y + k$ $+ 1$. Between 1991 and 1994, Shorey and Saradha published four papers dedicated to this topic and investigated various possibilities. They conducted a detailed study in this area and were gradually trying to generalize the problem. In 1994 and 1996, Shorey and his collaborators published two more papers in this line and established some important results.

Professor T. N. Shorey was quite active even during the first part of the twenty-first century and carried out important research work on various topics of number

---

[171] Section 3, (TNS. 32).

[172] Section 3, (18).

[173] Section 3, (TNS. 71).

theory. But because of the stipulated time frame, they are not being discussed here. He has published over 143 research papers, but his most important contributions are related to the applications of Baker's theory. Professor Shorey has also guided and mentored two well-known number theorists in India namely Shanta Laishram and Saranya Nair.

Appendix

[35] *Baker's method*: In transcendental number theory, Baker's theorem gives a lower bound for the absolute value of linear combinations of logarithms of algebraic numbers. Alan Baker solved a problem posed by Alexander Gelfond. Using that result and other earlier established results, he proved transcendence of many numbers, to derive effective bounds for the solution of some Diophantine equations, and to solve the class number problem of finding all imaginary quadratic fields with class number 1.

[36] *Laguerre polynomial*: These polynomials are named after Edmond Laguerre (1834–1886). They are the solutions of Laguerre's equation $xy'' + (1 - x)y' + ny = 0$, which is a second-order linear differential equation. This equation has nonsingular solutions only if n is a non-negative integer.

The next number theorist, whose contributions are discussed, was also a student of Prof. K. Ramachandra. He is R. Balasubramanian.

2.4.8 Ramachandran Balasubramanian

Ramachandran Balasubramanian (R. Balasubramanian) was born on March 15, 1951, in a village named Sikkal in Tamil Nadu (Fig. 2.11). He had his primary and secondary school education in erstwhile Tanjore (now, Thiruvarur) district. He completed his undergraduate studies (1967–1970) and postgraduate studies (1970–1972) in mathematics from Pushpam College, Poondy, in the same district. During his college years, he was inspired by his teacher Prof. V. Krishnamurthy.

Thereafter, in 1972, he moved to Bombay (now, Mumbai) and joined the premiere Tata Institute of Fundamental Research (TIFR) there. He went there to pursue his dream of doing research in mathematics. There he developed deep interest in "numbers". After successfully completing his first year's course work at TIFR, Balasubramanian started working under the supervision of the leading number theorist Prof. K. Ramachandra. Balasubramanian decided to work on analytic number theory for his topic of research as he felt the subject matter was more concrete than abstract. When R. Balasubramanian joined Prof. Ramachandra in 1973, Ramachandra's interest was mainly concentrated on the theory of Riemann zeta function.

Balasubramanian's Ph.D. Thesis, under the supervision of Prof. K. Ramachandra, dealt with an improvement of a theorem of Titchmarsh (1934) on the mean square of the Riemann zeta function and a theorem of Hardy (1914) on the gaps between zeros

Fig. 2.11 R.
Balasubramanian (b. 1951)

of the Riemann zeta function on the critical lines. These works were highly acclaimed and in 1978, fetched him the Young Scientist Award of Indian National Science Academy (INSA). After completing his Ph.D., he continued with his research work at TIFR and in 1981, he received an invitation to visit the Institute for Advanced Study (IAS), Princeton, USA. There he came in contact with many great and internationally famous mathematicians, including the Fields Medal winner Atle Selberg. While at Princeton, Dr. Balasubramanian collaborated with a well-known Indian number theorist who settled in Canada, namely M. Ram Murty. They jointly worked on the oscillations of the values of Ramanujan Tau Function. The next year, he moved to the University of Illinois, Urbana-Champaign, and worked there for a year. During his stay there, Dr. Balasubramanian in collaboration with J. B. Conrey and D. R. Heath-Brown derived an asymptotic formula for the mean square of the product of the Riemann zeta function and a Dirichlet polynomial [37]. This method led to an improvement of the lower estimate for the density of zeros of the Zeta function on the critical line.

After returning to India, Dr. Balasubramanian continued to work at the TIFR during 1983–1984. In 1985, he accepted an offer to work in a relatively new institute, namely, Institute of Mathematical Sciences (IMSc.) at Chennai. There at the initiative of Prof. C. S. Seshadri, a group on pure mathematics was set up.

During his early years at the TIFR, young researcher Balasubramanian had developed an interest in one of the 300-year-old classical problems in number theory, namely "Waring's problem". This problem has already been discussed in detail, earlier in this book. An important breakthrough was achieved by Dr. Balasubramanian on the case $k = 4$ of the said problem. Given a positive integer k, let $g(k)$ be the smallest number of kth powers which are needed to express a very positive integer as their sum. In 1986, Dr. Balasubramanian in collaboration with J. M. Deshouillers and F. Dress settled the difficult case $k = 4$ by proving $g(4) = 19$.

In 1990, as a mark of recognition of his brilliant research work on number theory, Dr. Balasubramanian was awarded the prestigious S. S. Bhatnagar Award. The same year, he became a Professor at IMSc and was also elected Fellow of the three national academies of India. Professor Balasubramanian has spent his entire work life as a faculty member and later as the Director of IMSc.

During 1990–2000, Prof. Balasubramanian in collaboration with another well-known Indian–Canadian number theorist, Prof. V. Kumar Murty, made many important contributions related to the Dirichlet L-function. From 1994–1996, Balasubramanian and K. Soundararajan together completely settled the "conjecture of Graham". In 1996, with N. Koblitz, he proved the Menezes–Okamoto–Vanstone algorithm for reducing the elliptic curve discrete log problem (DLP) to classical DLP over finite fields. This work guarantees that the elliptic curve based on cryptography is more secure than the one based on classical DLP. This is a remarkable result.

Professor R. Balasubramanian is one of India's leading number theorists and has been responsible for building up a vibrant school of research on number theory in India. Already, 11 students have obtained their Ph.D. degrees under his supervision. They are S. D. Adhikari, C. S. Yogananda, M. Vellammal, R. Padma, N. Amora, R. Venkataraman, M. Kulkarni, S. V. Nagaraj, D. S. Ramana, G. Prakash, and P. P. Pandey.

Appendix

[37] *Dirichlet polynomials*: In particular, they are partial sums of corresponding Dirichlet series. There exist inversion formulae for the Dirichlet series which give an integral expression of the Dirichlet polynomial by a sum of corresponding Dirichlet series.

[38] *Dirichlet L-function*: In number theory, Dirichlet L-series is a function of the form $L(s, X) = \sum X(n)/n^s$, where the summation over n is from 1 to infinity, X is a Dirichlet character and s is a complex variable with real part greater than 1. By analytic continuation, this function can be extended to a meromorphic function on the whole complex plane, and is then called Dirichlet L-function. It is mathematically denoted by $L(s, X)$.

2.4.9 Ayyadurai Sankaranarayanan

Ayyadurai Sankaranarayanan (A. Sankaranarayanan) was born in 1961 in Harikesa-vanallur village in Tirunelveli district of Tamil Nadu. He completed his graduation in 1982 from M. D. T. Hindu College affiliated to Madurai-Kamaraj University and did his M.Sc in mathematics from the same university in 1984. He completed his Ph.D. under the guidance of the famous number theorist Prof. K. Ramachandra of TIFR, Mumbai. He worked as a research scholar at TIFR from 1985–1989 and obtained his Ph.D. degree in 1991. The title of his thesis was "Some Problems in Analytic Number Theory". Professor Sankaranarayanan has spent his entire work life as a faculty member of the TIFR, Mumbai.

A brief account of his research contributions is given here. Identities involving the values of Riemann zeta function at even positive integers, Omega results of the Hurwitz zeta functions [38] on any vertical line in the critical strip, and improved Omega results for the Riemann zeta function on the left-side vertical lines nearer to the line 1 were established by him. In a sequence of papers, the results of Littlewood and Selberg pertaining to the Riemann zeta function were studied more generally by him. An unconditional result on the growth of the Riemann zeta function on certain horizontal lines in a bounded strip was established by Prof. Sankaranarayanan and this improved a result of Littlewood who obtained a weaker result of similar type assuming Riemann hypothesis. It may also be noted that quantitative results on the difference between consecutive zeros of quadratic zeta functions were established by him, which improved earlier known results. He also established Hardy's theorem for Zeta functions of quadratic forms. Error terms related to some important arithmetical functions were studied by Prof. Sankaranarayanan and interesting results were proved using real analytic methods.

Professor Sankaranarayanan is still actively involved in research on number theory. But because of the time constraints of the current book, his contributions made during the twenty-first century have not been fully mentioned.

Appendix

[39] *Hurwitz zeta function*: It is named after Adolf Hurwitz, and is one of the many Zeta functions. It is a generalization of the Riemann zeta function ζ (s). Hurwitz zeta function is formally defined for complex arguments s with Re $(s) > 1$ and q with Re $(q) > 0$ by the expression $\zeta(s, q) = \sum 1/(q + n)^s$ summation over n from 0 to infinity. The series is absolutely convergent for the given values of s and q, and can be extended to a meromorphic function defined for all $s \neq 1$.

2.4.10   Sukumar Das Adhikari

Sukumar Das Adhikari (S. D. Adhikari) was born in a village in Burdwan district of West Bengal in 1957. He completed his M.Sc. degree in mathematics from Burdwan University. He obtained his Ph.D. degree from Institute of Mathematical Sciences, Madras (now, Chennai), working under the guidance of Prof. R. Balasubramanian in 1991. The title of his thesis was "Some Omega Theorems and Related Questions in Number Theory". Professor S. D. Adhikari has spent most of his career as a senior faculty member at the Harishchandra Research Institute (HRI), Allahabad.

S. D. Adhikari has made some important contributions during the last part of the twentieth century. He is still an active and important researcher in the field of analytic number theory. But because of the time constraint of the current book, his contributions made during the twentieth century only are discussed.

A classical problem in number theory is related to the number $A_k(x)$ of integer lattice points; that is, points $(n_1, n_2, \ldots, n_k)$ where n_i's are integers, in a k-dimensional sphere of radius $\sqrt{x}$.

It is not difficult to observe that as x tends to infinity, $A_k(x)$ is roughly of the size $V_k(x)$, where $V_k(x)$ is the volume of the sphere. The problem here is to estimate and to determine the fluctutaions of the error term $P_k(x) = A_k(x) - V_k(x)$.

For $k \geq 5$, the problem has been completely solved and it is known that $P_k(x)$ is of the order of $(x^{k/2-1})$. For cases $k = 2, 3, 4$, the problem is still open. The case $k = 2$ is the "circle problem", which is one of the most famous unsolved problems in analytic number theory. Regarding the error term $P_4(x)$, Walfisz proved that $P_4(x) = O(x(\log x)^{2/3})$. That is, $|P_4(x)| \leq C(x(\log x)^{2/3})$, for a positive constant C.

In the other direction, from a theorem of Szegö, for $k = 3, 4$, one obtains that $P_k(x) = \Omega((x\log x)^{k-1/4})$. That is, $P_k(x)$ becomes less than $-C_1((x\log x)^{k-1/4})$ for a positive constant C_1, for infinitely many values of x tending to infinity.

For the case $k = 4$, Walfisz made the rather simple observation that the fact $\sigma(n) = \Omega(n\log\log n)$ can be employed to derive $P_4(x) = \Omega(x\log\log x)$. That is, the fluctuation of the order $(n\log\log n)$ can be employed to derive $P_4(x) = \Omega(x\log\log x)$. That is, the fluctuation is of the order $(n\log\log n)$ at least in the positive or in the negative direction. One observes that the corresponding result of Szegö cited above, though a bit more precise, leaves much room for improvement as it differs from that of Walfisz in the power of x itself. In the two research papers mentioned below, Adhikari and his collaborators managed to bridge this gap employing the method of Erdös and Shapiro. They are

- Adhikari, S. D., Balasubramanian, R., and Sankaranarayanan, A., "An Ω- result related to $r_4(n)$" [Sect. 2.3, (SDA. 1)];
- Adhikari, S. D. and Pétermann, Y. F. S. "Lattice points in ellipsoids" [Sect. 2.3, (SDA. 5)].

Adhikari has done further research in this area and references to his work are available in recent publications and also in the well-known book titled *Rational Number Theory in the* 20th *Century: From PNT to FLT* (by W. Narkiewicz), Springer. But these have all been published in the twenty-first century. So the time frame of the book prevents detailed discussions.

A question of Erdös, Gruber, and Hammer, regarding visibility of lattice points, was solved in a research paper by Adhikari and Balasubramanian published in 1996.

It may be noted that Prof. S. D. Adhikari's research works related to Ω-results and lattice points in ellipsoids have been acclaimed internationally and have made a notable impact on modern-day research in the area.

2.4.11   Dipendra Prasad

Born in 1960, Dipendra Prasad graduated from the St. Xavier's College, Mumbai, in 1980, and completed his master's degree in mathematics from IIT Kanpur. He got his Ph.D. degree from the University of Harvard, USA, in 1989, under the supervision of Prof. Benedict H. Gross. The title of his thesis was "Trilinear Forms for GL(2) of a Local Field and Epsilon-factors". He has spent a decade in his early career as a faculty member of the HRI, Allahabad, and at present he is a professor at TIFR, Mumbai. Professor Dipendra Prasad is one of India's noted number theorists. The contributions made by him, early in his career (during the twentieth century), are briefly discussed below.

Many problems in representation theory involve understanding how a representation of a group decomposes when restricted to a subgroup. Situations which involve multiplicity one in which either the trivial representation of the subgroup appears with multiplicity at most one, is especially useful. To cite a few examples, the spherical functions and the theory of Whittaker models depend on such a multiplicity one phenomenon. The Clebsch–Gordon theorem about tensor product of representations of SU(2)—again involving multiplicity one—has been very useful both in physics and mathematics. Many of Prof. Dipendra Prasad's works have been about finding such multiplicity one situations for infinite-dimensional representations of real and p-adic groups. The results have been expressed in terms of the arithmetic information which goes in parameterizing representations, the so-called "Langlands parameters". In particular, the Clebsch–Gordon theorem was generalized by Prasad for infinite-dimensional representations of real and p-adic GL(2). Several of his papers, written in collaboration with B. H. Gross, point out to the importance of the so-called epsilon factor in these branching laws. There are many parallels between global period integrals expressed in many situations as a special value of L-functions, and local branching laws expressed in terms of epsilon factors.

His research work on automorphic representations [39] is widely acclaimed. In 2012, the Gan–Gross–Prasad conjecture was made and it has been considered as a landmark contribution. But since the book is only concerned with the twentieth century contributions, the detailed discussions are not done.

Appendix

[40] *Automorphic representation*: Inside an L^2-space for a quotient of the adelic form of G, an automorphic representation is a representation that is an infinite tensor product of representations of p-adic groups, with specific enveloping algebra representations for the infinite prime (s).

2.4.12　K. Soundararajan

The unusual success story of a young promising Indian mathematician K. Soundararajan would complete this brief write-up on noted number theorists of this country who made notable contributions during the twentieth century.

While studying in the 8th standard at a school in Chennai in 1985, K. Soundararajan was spotted as a precocious student of mathematics and was brought to Matscience (new name, Institute of Mathematical Sciences, Chennai). He regularly visited the Institute and studied in its excellent library. Subsequently, he came in close contact with Prof. R. Balasubramanian, Prof. C. S. Seshadri F.R.S., and other faculty members who worked there. The same year, Soundararajan discovered that the only four-digit number having two repeated adjacent digits and which itself was the square of a number with repeated digits was $7744 = 88^2$. Encouraged by his mentors in Matscience, his parents, and teachers in school, he appeared in the All India Olympiad Tests in mathematics from 1987 to 1991. He stood first in India in those tests. While still studying in school in Madras, he achieved a lot of success and won many international laurels for his mathematical achievements. After finishing school, he went to the USA to pursue studies in higher mathematics. During the summer programmes, he continued working with Prof. R. Balasubramanian and published several research papers on number theory. He went on to complete his Ph.D. from Princeton University, Princeton, under the supervision of Prof. Peter Sarnack. He is a well-known number theorist and like many other Indian mathematicians, settled in the USA.

This brief paragraph has been devoted to K. Soundararajan, because in a sense he was discovered and nurtured by the faculty members of Matscience in the early formative years of his life.

References

1. Ramanujan, S.: Some properties of Bernoulli's numbers. J. Indian Math. Soc. **3**, 219–234 (1911)
2. Ramanujan, S.: On certain arithmetical functions. Trans. Cambridge Philos. Soc. **22**(9), 159–184 (1916)
3. Ramanujan, S.: Some properties of $p(n)$, the number of partitions of n. Proc. Cambridge Philos. Soc. **19**, 207–210 (1919)
4. Ramanujan, S.: Congruence properties of partitions. Proc. Lond. Math. Soc. **2**, 18 (1920)
5. Ramanujan, S.: Congruence properties of partitions. Mathematische Zeitschrift, 147–153 (1921)
6. Ramanujan, S.: Modular equations and approximations to π. Q. J. Math. **45**, 350–372 (1914)
7. Ramanujan, S., Hardy, G.H.: Une formulae asymptotique pour le nombre des partitions de n. Comptes Rendus (1917)
8. Ramanujan S. (with G. H. Hardy) "Asymptotic formulae in combinatory analysis". *Proceedings of the London Mathematical Society*, 2. 17(1918)75–115.
9. Ramanujan, S., Hardy, G.H.: Asymptotic formulae for the distribution of integers of various types. Proc. Lond. Math. Soc. **2**(16), 112–132 (1917)
10. Ramanujan, S., Hardy, G.H.: Asymptotic formulae in combinatory analysis. Proc. Lond. Math. Soc. **2**, 16 (1917)
11. Ramanujan, S.: Proof of certain identities in combinatory analysis. Proc. Cambridge Philos. Soc. **19**, 214–216 (1919)
12. Ramanujan, S.: Highly composite numbers. Proc. Lond. Math. Soc. **2**(14), 347–409 (1915)
13. Ramanujan, S., Hardy, G.H.: The normal number of prime factors of a number n. Q. J. Math. **48**, 76–92 (1917)
14. Ananda Rau, K.: The infinite product for $(s-1)\zeta(s)$. Mathematische Zeitschrift **20**, 156–164 (1924)
15. Ananda Rau, K.: On the boundary behaviour of elliptic modular functions. Acta Mathematica **52**, 143–168 (1929)
16. Ananda Rau, K.: Additional note on the boundary behaviour of elliptic modular functions. Acta Mathematica **53**, 77–86 (1929)

17. Ananda Rau, K.: On the representation of a number as the sum of an even number of squares. J. Madras Univ. B **24**, 61–89 (1954)
18. Ananda Rau, K.: On the behaviour of elliptic theta functions near the line of singularities. J. Indian Math. Soc. **20**, 148–156 (1933)
19. Ananda Rau, K.: On the summation of singular series associated with certain quadratic forms I. J. Indian Math. Soc. (N. S.) **23**, 65–96 (1959)
20. Ananda Rau, K.: Application of modular equations to some quadratic forms. J. Indian Math. Soc. (N. S.) Jubilee Issue, **24**, 77–130 (1960)
21. Ananda Rau, K.: Relation between sums of singular series associated with certain quadratic forms. J. Madras Univ. B **31**, 7–10 (1961)
22. Ananda Rau, K.: On the summation of singular series associated with certain quadratic forms II. J. Indian Math. Soc. (N. S.) **25**, 173–195 (1961)
23. Pillai, S.S.: On some diophantine equations. J. Indian Math. Soc. **18**, 291–295 (1930)
24. Pillai, S.S.: On the inequality $0 < a^x - b^y < n$. J. Indian Math. Soc. **19**, 1–11 (1931)
25. Pillai, S.S.: On $a^x - b^y = c$. J. Indian Math. Soc. (N. S) **II**, 119–122 (1936)
26. Pillai, S.S.: Correction to the paper 'On $a^x - b^y = c$'. J. Indian Math. Soc. (N. S) **II**, 215 (1936)
27. Pillai, S.S.: On numbers of the form $2^a 3^b$–I. Proc. Indian Acad. Sci. Sect. A **XV**, 128–132 (1942)
28. Pillai, S.S., George, A.: On numbers of the form $2^a 3^b$–II. Proc. Indian Acad. Sci. Sect. A **XV**, 133–134 (1942)
29. Pillai, S.S., Chowla, S.: On the error terms in some asymptotic formulae in the theory of numbers I. J. Lond. Math. Soc. **5**, 95–101 (1930)
30. Pillai, S.S., Chowla, S.: On the error terms in some asymptotic formulae in the theory of numbers II. J. Indian Math. Soc. **18**, 181–184 (1930)
31. Pillai, S.S., Chowla, S.: Hypothesis K of Hardy and Littlewood. Mathematische Zeitschrift **41**, 537–540 (1936)
32. Vijayaraghavan, T.: Periodic simple continued fractions. Proc. Lond. Math. Soc. **26**, 403–414 (1927)
33. Pillai, S.S.: On a linear diophantine equation. Proc. Indian Acad. Sci. Sect. A **XII**, 199–201 (1940)
34. Vijayaraghavan, T.: A note on diophantine approximation. J. Lond. Math. Soc. **2**, 13–17 (1927)
35. Vijayaraghavan, T.: On the irrationality of a certain decimal. Proc. Indian Acad. Sci. A **10**, 341 (1939)
36. Vijayaraghavan, T.: On decimals of irrational numbers. Proc. Indian Acad. Sci. A **12**, 20 (1940)
37. Vijayaraghavan, T.: On the fractional parts of the powers of a number, I. J. Lond. Math. Soc. **15**, 159–160 (1940)
38. Vijayaraghavan, T.: On the fractional parts of the powers of a number III. J. Lond. Math. Soc. **17**, 137–138 (1942)
39. Vijayaraghavan, T.: On the fractional parts of the powers of a number, IV. J. Indian Math. So. **12**, 33–39 (1948)

Chapter 3
Impact of the Research of the Indian Number Theorists on Modern Mathematics

As soon as one sits down to write about the impact of number theory-related research of Indian mathematicians on modern mathematics, the first name that naturally comes to one's mind is that of Srinivasa Ramanujan. Here, it would be quite apt to quote Freeman J. Dyson who wrote:

> The seeds from Ramanujan's garden have been blowing on the wind and have been sprouting all over the landscape.

The total volume of Ramanujan's work is so vast that a selective approach is imperative. The first discussion relates to Ramanujan's much acclaimed work on *partition function congruences.*

Partition function has already been defined earlier. For a positive integer n, the partition function $p(n)$ is the number of partitions of n into positive integral parts. In a partition, the parts are not necessarily distinct and the order in which the parts are arranged is not relevant. From MacMahon's table for $p(n)$, Ramanujan in his paper titled "Some properties of $p(n)$, the number of partitions of n",[1] conjectured the following congruence properties of $p(n)$:

$$\text{If } 24m \equiv 1 \left(\text{mod } 5^a 7^b 11^c\right), \text{ then } p(m) \equiv 0 \left(\text{mod } 5^a 7^b 11^c\right). \tag{3.1}$$

In the paper mentioned above, Ramanujan also proved the particular cases of $p(5n + 4) \equiv 0 \pmod 5$ and $p(7n + 5) \equiv 0 \pmod 7$.

Hardy used materials from Ramanujan's "Unpublished manuscript on the partition and tau functions"[2] and published a paper titled "Congruence properties of partitions",[3] posthumously in the name of Ramanujan. There, the following congruence was proved: $p(11n + 6) \equiv 6 \pmod{11}$. In a paper titled "Ramanujan's unpublished

---

[1] Section 1 (SR 20).

[2] *The Lost Notebook and other Unpublished Papers.* New Delhi: Narosa. 1988.

[3] Section 1 (SR 23).

© The Author(s), under exclusive license to Springer Nature Singapore Pte Ltd. 2020
P. Mukherji, *Research Schools on Number Theory in India,*
https://doi.org/10.1007/978-981-15-9620-9_3

manuscript on the partition and tau functions with proofs and commentary",[4] Bruce. C. Berndt and Ken Ono have clarified, corrected, and given an excellent commentary on the posthumous publication. Incidentally, as has been already discussed earlier in Chap. 2, from the extended tables of $p(n)$ prepared by Hansraj Gupta, it was noticed by S. Chowla in 1934 that $p(243)$ was not divisible by 7^3. This led to the suitable modification of Ramanujan's first conjecture (1) mentioned earlier by G. N. Watson.

In the unpublished manuscript mentioned earlier, Ramanujan suggested some line of proof for the congruence $p(121n - 5) \equiv 0 \pmod{121}$. J. M. Rushforth in his paper titled "Congruence properties of the partition function and associated functions"[5] completed the proof. He did not exactly follow the method indicated by Ramanujan and used his own technique. In his two papers entitled "Ramanujan identities involving the partition function for the moduli 11^a"[6] and "Proof of Ramanujan's partition congruence for the moduli 11^3",[7] D. H. Lehmer followed Rademacher and developing a method different from that of Ramanujan–Watson, and proved the modified conjecture for 11^3 and 11^4. Later, A. O. L. Atkin in his paper titled "Proof of a conjecture of Ramanujan"[8] for the first time was successful in proving Watson's modified conjecture for 11^n for all positive integers n. A large volume of research was done on congruences of partition functions modulo powers of 5, 7, and 11. Compared to 5 and 7, the proofs of congruence modulo 11 were more difficult. In a paper titled "An elementary proof of $p(11n + 6) \equiv 0 \pmod{11}$",[9] L. Winquist gave the elementary proof of $p(11n + 6) \equiv 0 \pmod{11}$. In the nineties of the twentieth century, F. Garavan and D. Stanton[10] and F. Garavan, D. Kim, and D. Stanton[11] did a considerable amount of work in this area. Following his predecessors, M. D. Hirschhorn in his paper entitled "Ramanujan's partition congruence"[12] gave simplified versions of proofs of all the three congruences:

$$P(5n + 4) \equiv 0 \pmod 5, \; p(7n + 5) \equiv 0 \pmod 7 \text{ and } p(11n + 6) \equiv 0 \pmod{11}.$$

Ramanujan's conjectures of partition functions have been a source of the quest for many famous mathematicians throughout the twentieth century. Apart from the mathematicians discussed above, F. J. Dyson, G. E. Andrews, B. C. Berndt, K. Ono, A. Schinzel, E. Wirsing, M. Newman, P. Erdö, P. Swinnerton-Dyer, and many others have carried out remarkable research in this area giving a new dimension to the conjectures and also by utilizing them to prove other problems in other branches of

[4](Maartea, 1998), Sém Lothar, Combin, 42, (1999).

[5]*Proceedings of the Cambridge Philosophical Society* 48 (1952), 402–413.

[6]*American Journal of Mathematics* 65 (1943), 492–520.

[7]*Proceedings of the American Mathematical Society* 1 (1950), 172–181.

[8]*Glasgow Mathematical Journal* 8 (1967), 14–32.

[9]*Journal Combinatorial Theory* 6 (1969), 56–59.

[10]*Math. Comp.* 55 (1990), 299–311.

[11]*Invent. Math.* 101 (1990), 1–17.

[12]*Discrete Mathematics* 131 (1994), 351–355.

mathematics. So, the impact of Ramanujan's conjectures on partition functions is self-evident.

The next topic of discussion is the impact of *Ramanujan's τ-function*. This function has already been discussed in detail in Chap. 2. Here, our main focus is to use this function as a good illustration of Ramanujan's insight and ingenuity in finding new and interesting areas of mathematics that remained untouched by other mathematicians of repute. It will also be shown how his conjectures in this area have made a tremendous impact on modern mathematics.

From historical records, it appears that Ramanujan was the first mathematician to show any interest in the coefficient $\tau(n)$. He found and investigated three main types of properties, as stated below:

(i) $\tau(mn) = \tau(m) \, \tau(n)$, for $(m \cdot n) = 1$
(ii) $T(p^{r+1}) = \tau(p) \, \tau(p^r) - p^{11} \tau(p^{r-1})$ [p is prime, $r \geq 1$]
(iii) Order of magnitude of $\tau(n)$, and in particular the conjecture $|\tau(p)| \leq 2p^{11/2}$ (p prime).

Ramanujan had discussed the abovementioned properties in detail in a classic paper titled "On certain arithmetical functions".[13]

In 1917, L. J. Mordell in his paper titled "On Ramanujan's empirical expansions of modular functions"[14] proved the first two conjectures. E. Hecke[15] in these two research papers generalized the proofs given by Mordell and clearly explained the theory behind it. R. A. Rankin in his two publications[16] was able to show that $\tau(n) = O(n^{11/2 + 3/10})$ by using analytic methods. That was the best result obtained till then.

It may be recalled that Ramanujan made his conjectures in 1916. A whole galaxy of mathematicians was unsuccessful for nearly six decades till 1974. The validity of the third conjecture was established by the outstanding French mathematician P. Deligne. He used very sophisticated and powerful techniques from algebraic geometry to prove the conjecture. Deligne was awarded the Fields Medal for this work.

The third and perhaps the most remarkable impact of Ramanujan's work relates to his conjecture on the order of magnitude of Fourier coefficients of cusp forms. It has made a tremendous impact on present-day mathematics and thereby propelled rapid advances in the discipline. If one goes back a little in the history of mathematics, it may be noted that under the influence of E. Artin, André Weil in his paper titled "Number of solutions of equations in finite fields"[17] introduced the concept of Zeta function of an algebraic type over a finite field. His conjecture was that the said Zeta function was a rational function satisfying an appropriate functional equation and was also an analogue of the Riemann hypothesis. These are known as the famous *Weil conjectures*.

---

[13] Section 1 (SR 10).

[14] *Proceedings of the Cambridge Philosophical Society* 19 (1920), 117–124.

[15] *Math. Ann.* 114 (1937); *Math. Ann.* 114 (1937), 316–351.

[16] *Proceedings of the Cambridge Philosophical Society* 35 (1939), 357–372; *Proceedings of the Cambridge Philosophical Society* 36 (1940), 150–151.

[17] *Bulletin of the American Mathematical Society* 55 (1949), 497–508.

Influenced and motivated by Ramanujan, A. Selberg in his paper titled "On the estimation of Fourier coefficients of modular forms",[18] and U. V. Linnik is his paper titled "Additive problems and eigenvalues of modular operators"[19] independently conjectured that $\tau(p) = O_\varepsilon(p^{11/2+\varepsilon})$, for every $\varepsilon > 0$. Not going into the detailed mathematics, it may be noted that the Sato–Tate conjecture followed J. P. Serre's proposition. In his paper titled "On the Sato–Tate conjecture",[20] V. Kumar Murty established some important results in this connection. In 1943, D. H. Lehmer in his paper titled "Ramanujan's function $\tau(n)$",[21] by using empirical evidence conjectured that $\tau(p) \neq 0$ for every p. As a part of a more general conjecture on Fourier coefficients of cusp forms, S. Lang and H. Trotter in their article titled "Frobenius distributions in GL_2 extensions"[22] made another conjecture. M. Ram Murthy, V. Kumar Murthy, and T. N. Shorey in their paper entitled "Odd values of the Ramanujan τ-function"[23] proved a variant of the conjecture of Lang and Trotter. The conjectures mentioned above and their proofs which have not been discussed here because of the heavy mathematics involved are all offshoots of the pathbreaking ideas as conceived by Srinivasa Ramanujan. In this context, quoting the noted number theorist M. Ram Murty is relevant.

He wrote in the context of various important conjectures and their proofs which emerged as a consequence of Ramanujan's investigations on τ-functions:

> These results reveal to some extent the arithmetical significance of the τ-function and Fourier coefficients of cusp forms in general. The story is not yet over. The future awaits for some deeper connections between non-Abelian class field theory and Fourier coefficients of cusp forms.

3.1   Rogers–Ramanujan Identities

In the entire theory of partitions and q-series, the Rogers–Ramanujan identities are the most simple, elegant, and deep results. The statement of the first identity is.

> The number of partitions of an integer into parts differing by at least 2 equals the number of partitions of that integer into parts which when divided by 5 have remainder 1 or 4.

The statement of the second identity is also similar. This pair of identities was independently discovered by L. J. Rogers in 1894 and by S. Ramanujan in 1910. But both of them had only stated the analytic form of the identities. I. Schur of Germany had independently worked and successfully proved the identities and was

---

[18] *Proc. Symposia in Pure Maths.* VIII (1965), 1–15, American Mathematical Society, Providence.

[19] *Proceedings of the International Congress of Mathematics*, Stockholm, (1962), 270–284.

[20] *Number Theory Related to Fermat's Last Theorem*, Ed. N. Koblitz (1982), 195–205. Birhauser-Verlag, Boston.

[21] *Duke Mathematical Journal* 10 (1943), 483–492.

[22] *Springer Lecture Notes in Mathematics* 504 (1976), Springer-Verlag.

[23] *Bulletin Soc. Math. France*, (1987), 115.

the first mathematician who realized their combinatorial significance. He was able to obtain the next level partition theorem, which is presently known as *Schur's partition theorem*.

As has been discussed elsewhere, Ramanujan was motivated to work on identities while studying an infinite continued fraction. He had the insight to realize the importance of this continued fraction in the theory of modular forms. This continued fraction can be obtained as the ratio of two Rogers–Ramanujan series. Ramanujan's continued fraction is one of the most fundamental objects in the theory of modular forms. Starting from the beginning of the 1960s for around two-and-a-half decades, a substantial amount of work has been done in this area.

From the historical perspective, it may be important to note that these beautiful identities appear in Chap. 16 of Ramanujan's *Second Notebook*. The chapter titled "Theta functions and q-series" has been thoroughly analyzed and edited with great care by C. Adiga, B. C. Berndt, S. Bhargava, and G. N. Watson.

Ramanujan had sent 40 identities of such types to the English mathematician L. J. Rogers, but the proofs were not there. These identities involved modular functions for congruence subgroups. Rogers was able to prove nine of them and published the proofs. G. N. Watson proved several of these identities. G. N. Watson, W. N. Bailey, B. Gordon, L. J. Rogers, L. J. Slater, and G. E. Andrews did a considerable amount of work toward the generalization of the Rogers–Ramanujan identities.

Rogers–Ramanujan identities have occurred in various settings. They occur in the study of Euclidean Lie algebras. In this context, searching for proofs led to a deeper understanding of some of the representation theories of the Lie algebras. In 1980, in R. J. Baxter's solution to the hard hexagon model in statistical mechanics, the said identities appeared naturally. Subsequently, Andrews and Baxter were able to work out a complete set of solutions. More recently, Prof. Barry McCoy in collaboration with Alexander Berkovich, Anne Schilling, and Ole Warner were successful in obtaining new extensions of various Rogers–Ramanujan-type identities by studying models in the conformal field theory in physics.

Andrews and Bressoud were able to detect that there was a pattern among the coefficients of certain Rogers–Ramanujan-type products having value zero. Professors Krishnaswami Alladi and B. Gordon have recently extended these results to general Rogers–Ramanujan-type products. According to these mathematicians, there is a scope for more work in that area.

3.2 Hypergeometric Series

In Ramanujan's *Second Notebook*, Chaps. 10 and 11, one can find a number of results related to the hypergeometric series. Ramanujan made asymptotic expansions for such series and related functions. G. N. Watson, R. Askey, and B. Berndt have carried out detailed discussions on these types of work. Without going into the mathematical details, it may be stated that classical basic hypergeometric functions of one variable (that is q-series) have been usefully and effectively applied to diverse areas of pure as well as applied mathematics. In recent times, q-series have given rise to new and important developments in physics, Lie algebra, transcendental number theory, and statistics.

As already stated at the outset, the total volume of Ramanujan's work is vast and the impact of his contributions is immense. Solving his unsolved problems and conjectures is one part of it. But the numerous questions raised by him have triggered fresh new questions and opened up fresh avenues of investigations which include the emergence of completely new disciplines. There lies the immense importance of the work of this iconic genius.

3.3 Waring's Problem

The next major research contributions by some noted Indian number theorists, which have a great impact on traditional number theory, relates to a problem called *Waring's problem*. A brief historical development of the problem and the impact made by the contributions of Indian number theorists will be briefly discussed here.

E. Waring (1734–1798) considered the possibility of representing positive integers as the sum of powers. In this respect, he did not give any proof, but formulated an assertion that came to be known as "Waring's problem". He wrote:

> Every integer is a sum of two, three, ..., nine cubes, every integer is also the square of a square or up to the sum of nineteen such; and so forth. Similar laws may be affirmed for the correspondingly defined numbers or quantities of any degree.

$g(k)$ may be denoted as the smallest number of terms sufficient to represent any positive integer as the sum of kth powers of non-negative integers. Thus, Waring's assertion means that $g(3) = 9$, $g(4) = 19$, etc. It may be said that $g(k)$ is finite for every $k \geq 2$.

In the case of squares, Descartes, Fermat, and others had formulated their theorem. L. Euler tried several times to prove it but met with no success. C. Bachet (1581–1638) had tried to prove the problem for $k = 2$ but ultimately the famous French mathematician J. L. Lagrange established that $g(2) = 4$ in 1770.

Waring himself had checked for numbers up to 3000 that every number was the sum of nine cubes. C. G. J. Jacobi carried out the exercise for numbers up to 12,000. As a consequence of his findings, he conjectured that all large integers are the sum of at most five cubes, and every integer larger than 8042 is the sum of at most six positive cubes. However, these are still unproved and remain open conjectures. The present modified version of Waring's problem is stated as follows:

> For each natural number k, there exists a natural number n_k such that every natural number is a sum of at most n_k k-th powers of natural numbers.

The smallest of such numbers n_k is by convention denoted by $g(k)$. This is called Waring's constant for the exponent k.

So there were actually two problems to be solved. Firstly, the existence of $g(k)$ had to be proved. Secondly, $g(k)$ had to be determined. During the middle of the nineteenth century, J. Liouville started investigating the problem of fourth powers. He obtained $g(4) \leq 53$. Further research was carried out by various mathematicians and during the nineteenth century, the values of $g(3)$, $g(4)$, and $g(5)$ were found to be finite, and explicit bounds for $g(3)$, $g(4)$, and $g(5)$ were established by E. Maillet and E'. Lucas. In 1907, E. Maillet proved the existence of $g(7)$. In 1909, I. Schur for the first time proved the existence of $g(10)$. A. Fleck, E. Landau, A. J. R. Wieferich, and many other mathematicians worked and established the bounds for $g(4)$, $g(5)$, and $g(7)$. In 1909, Wieferich was the first mathematician who determined $g(k)$, for k different from 1 and 2 and showed that $g(3) = 9$. Kempner also found the same value in 1912. In 1909, D. Hilbert gave the complete proof of the existence of the number $g(k)$ for arbitrary k and published it. This implied that $g(k)$ is finite for every k. He was successful in establishing the essential part of Waring's conjecture. However, Hilbert's proof was rather difficult and it involved the evaluation of complicated multi-dimensional integrals. His method was ineffective because it showed the finiteness of $g(k)$ but was not able to provide any upper bound of $g(k)$ in terms of k.

Hardy and Ramanujan, in their paper entitled "Asymptotic formulae in combinatory analysis",[24] while investigating the partition functions $p(n)$ developed a new analytic method for obtaining the asymptotic behavior of an arithmetical function. The method known as the "circle method" is a beautiful idea used in additive number theory. Later after the demise of Ramanujan, Hardy and Littlewood further modified and refined it and they applied the method to Waring's problem. They stated that the minimal non-negative kth powers needed to represent every sufficiently large integer are denoted by $G(k)$. Using the "circle method", Hardy and Littlewood in 1922 obtained the bound $G(k) < \text{or} = (k - 2) . 2^{k-1} + 5$ for any $k \geq 1$.

From the work of Hardy and Littlewood, it became clear that the problem of improving upon their upper bound $G(k)$ was the central theme for further progress on Waring's problem. It may be noted that earlier in 1908, A. Hurwitz and E. Maillet had been successful in proving the lower bound $G(k) \geq 1 + k$. This inequality still holds. I. M. Vinogradov published a paper in 1928, where he introduced major technical simplifications into the method of Hardy and Littlewood. This new modified method helped Vinogradov to recover the bound established by Hardy and Littlewood with a lot of brevity and simplicity. In 1936, Vinogradov in another research paper improved Weyl's sum and used it to show that the asymptotic formula for the number of representations of an integer as the sum of Skth powers holds for $S > 10k^3 \log k$. In 1938, Vinogradov reduced his bound to $G(k) \leq 4k (\log k + 2 \log \log k + 3)$ (for $k \geq 800$).

With so much work done by Vinogradov, between 1935 and 1940, an intense amount of research was done on the determination of $g(k)$. S. S. Pillai of India played a leading role in this. Independently, L. E. Dickson also played a very important role. In 1936, Dickson was able to establish the truth of the ideal Waring theorem for $11 \leq k \leq 15$ and $k = 17$. He was able to reduce the value of $g(k)$ for most k by establishing

---

[24] Section 1 (SR 14).

a useful theorem. The theorem helped to prove the truth of the ideal Waring theorem for $7 \leq k \leq 180$. In the same year, H. S. Zuckerman independently established the result for $15 \leq k \leq 20$.

In 1936–1937, S. S. Pillai independently established the first part of Dickson's theorem. In 1939, this was used by S. Chowla in his paper titled "A Remark on $g(n)$"[25] to prove that the ideal Waring's theorem holds for a set of integers of upper density 1. In 1936, S. S. Pillai showed that $g(6) \leq 104$ and in 1940, following the work of Van Der Corput on Weyl's inequality, he obtained $g(6) = 73$. Earlier, he had completed the determination of $g(7)$ by showing that $g(7) = 143$. These results of Pillai were probably the most important mathematical achievements in India of his time. This also made a tremendous impact on the international research of Waring's problem.

The remaining cases of small integers k, $k = 4$, and $k = 5$ were settled much later. In 1964, J. R. Chen showed that $g(5) = 37$. Another noted Indian number theorist R. Balasubramanian in collaboration with J. M. Deshouillers and F. Dress showed that $g(4) = 19$, which means that every natural number is a sum of no more than 19 biquadrates, that is the fourth powers of natural numbers. This was an original assertion of Waring. Apart from these major contributions by the two leading Indian number theorists S. S. Pillai and R. Balasubramanian, many other Indian number theorists have made important contributions in the field. They include S. Chowla, F. C. Auluck, and C. P. Ramanujam. Ramanujam's work is related to Waring's problem for number fields.

The next topic is on the "geometry of numbers". This is an area of number theory, which was initiated by Minkowski in his work on discriminants and class numbers of number fields. The Panjab School of Number Theory at Chandigarh under the leadership of R. P. Bambah has made valuable contributions in this area. Some of their results have been acclaimed internationally and have made a notable impact on modern number theory.

Historically speaking, R. P. Bambah is the first Indian number theorist who has worked in the area of the geometry of numbers. His research work submitted as a thesis for the Ph.D. degree to the University of Cambridge was carried out under the supervision of the internationally reputed number theorist L. G. Mordell. In his thesis titled "Some results in the geometry of numbers", Bambah developed a technique for finding out the critical determinant of non-convex star regions with hexagonal symmetry. He also extended some of the results which had earlier been established by Prof. Mordell. In 1952, Bambah in collaboration with C. A. Rogers and K. F. Roth published a series of two papers[26] dealing with lattice coverings. After his return to India in 1951, Bambah developed the theory of coverings. The work that he carried out in the fifties and the sixties of the twentieth century led to the study of finite coverings. After the partition of India, the University of Panjab was shifted from Lahore (now in Pakistan) and the new campus of the university was established at

---

[25] Section 2 (SC 96).
[26] Section 2 (RPB 23, 24).

Chandigarh. As already mentioned earlier, Bambah and Hansraj Gupta jointly set up the "School of Number Theory" in the new campus of the Panjab University.

In "geometry of numbers", there is a classical conjecture, on the product of n non-homogeneous linear forms, which was made by H. Minkowski (1864–1909). Known as *Minkowski's conjecture*, it is one of the most challenging problems in the field.

Minkowski himself in 1899, and later Mordell in 1928, 1941, 1953, Landau in 1931, Perron in 1938, Pall in 1943, Macbeth in 1948, 1961, Sawyer in 1948, Cassels in 1953, and some other number theorists had proved the conjecture for $n = 2$. The first proof of the conjecture for $n = 3$ was given by Remak in 1923. A simplified proof was given by Davenport in 1939. Birch and Swinnerton–Dyer in 1956 and Narzullaev in 1968 gave proofs on different lines. The Remak–Davenport method consists of two parts. For $n = 4$, Dyson in 1948 proved both the parts and hence the conjecture. For proving the first part, he used powerful tools from algebraic topology. Bambah and Woods in 1974[27] gave an elementary proof of the first part. In 1973, Skubenko gave a proof for the first part for $n = 5$.

In 1980, Bambah in collaboration with A. C. Woods in their paper titled "Minkowski's conjecture for $n = 5$; a theorem of Skubenko"[28] had proved the first part of the said conjecture for $n = 5$. Since Woods had already proved the second part, the proof for Minkowski's conjecture for $n = 5$ was complete.

Since the time frame of this book is the twentieth century, discussions beyond that period are not taken up. However, for the sake of completeness, it may be noted that the number theorists of the Panjab School have carried out exhaustive research on this topic in the twenty-first century as well.

Another important conjecture in the field of "geometry of numbers" is *Watson's conjecture*. This conjecture concerns non-homogeneous real indefinite quadratic forms. The researchers from the Panjab School have made important contributions in the field of quadratic forms which led to the proof of Watson's conjecture in 1962. In the work of her Ph.D. thesis, Madhu Raka in collaboration with R. J. Hans-Gill[29] determined the relevant constants $C_{n,\sigma}$ for $n = 5$ and all signatures. In her papers,[30] she also obtained $C_{n,\sigma}$ for signatures $\pm 1, \pm 2, \pm 3, \pm 4$, and all n. Several other mathematicians like Davenport, Birch, and Dumir have worked toward the solution of this conjecture. Since the conjectured value of $C_{n,\sigma}$ depended on the class of σ modulo 8, this was an important and major contribution toward the proof which was completed by Dumir, Hans-Gill, and Woods in 1994.

The third important conjecture, in the area of "geometry of numbers" is the *Oppenheim conjecture*. In Diophantine approximation, the Oppenheim conjecture concerns representations of numbers by real quadratic forms in several variables. It was first formulated by Alexander Oppenheim in 1929 and later the conjectured property was further strengthened by Harold Davenport and Oppenheim. Initially, researchers on

---

[27] Section 2 (RPB 53).

[28] Section 2 (RPB 56).

[29] Section 2 (MR 1, 2, 3, 4).

[30] Section 2 (MR 7, 8, 9, 10).

this problem took the number n variables to be large, and applied a version of the Harry–Littlewood circle method. Using methods arising from ergodic theory and the study of discrete subgroups of semisimple Lie groups, Grigory Margulis settled the conjecture in 1986–1987. In this context, it may be relevant to mention Margulis actually proved a conjecture of M. S. Raghunathan of TIFR on the closure of orbits of unipotent flows on $\Gamma \backslash G$, where G is a Lie group, and Γ is an arithmetic subgroup. S. G. Dani has also made substantial contributions in this area. Much work on Oppenheim conjecture has been done by Bambah, Raghavan, and Ramanathan, Raghunathan, and Dani. It will be discussed later.

There is another theorem related to the values of quadratic forms due to H. Blaney which has been stated by him in his paper titled "Indefinite quadratic forms in n variables".[31] Blaney in his theorem had mentioned the existence of a constant $\Gamma_{r,s}$ depending only on the signature (r, s) of Q (where Q is an indefinite quadratic form in n variables and of discriminant $D \neq 0$) at infinity and not on Q itself. Much work has been done by Bambah, Hans-Gill, Madhu Raka, and Urmila Rani all from the Panjab School of Chandigarh. In a survey paper authored by Bambah, Dumir, and Hans-Gill and titled "Non-homogeneous problems: conjectures of Minkowski and Watson",[32] a detailed account has been given about the values of $\Gamma_{r,s}$. Some notable examples are given below:

(i) $\Gamma_{1,1} = 4$ (calculated by Davenport and Heilbronn)
(ii) $\Gamma_{2,1} = 4$ (Blaney)
(iii) $\Gamma_{1,2} = 8$, $\Gamma_{3,1} = 16/3$, $\Gamma_{2,2} = 16$ (all determined by Dumir)
(iv) $\Gamma_{1,3} = 16$ (due to Dumir and Hans-Gill). The only constant not determined so far is $\Gamma_{5,3}$, for which Madhu Raka and her collaborators have obtained the upper bound 12 in 1997. The expected value is 8.

In a joint work, Madhu Raka and her scholar co-workers have also determined several successive extreme values for ternary and quaternary quadratic forms related to $C_{n,\sigma}$ and $\Gamma_{r,s}$. In the process, they detected a mistake of Rieger (1976) and gave a correct proof in 1994, in a paper written jointly with Urmila Rani. In 1991, Madhu Raka also detected a mistake in the work of Barnes and Swinnerton–Dyer (1954) on binary non-homogeneous quadratic forms, and working jointly with Grover gave a correct proof and used those results for obtaining second non-homogeneous minima for non-zero ternary and quaternary forms and published these results during 1994–2001.

Sudesh K. Khanduja of the Panjab School has contributed several important research papers around Hensel's lemma. Another important contribution of this noted number theorist from Chandigarh relates to the "Brauer–Siegel theorem". This famous theorem in algebraic number theory tells how the various basic invariants of a number field such as a class number, regulator, and the discriminant vary as

---

[31] *Journal of the London Mathematical Society* 23, (1948), 153–160.
[32] *Number Theory*, 15–41, *Trends Math.*, Birkhuser, Basel 2000.

the number field is varied. S. K. Khanduja (known as Sudesh K. Gogia before her marriage) and her Ph.D. supervisor I. S. Luthar proved the analogue for function fields.

There has been a number of publications by A. R. Rajwade, M. K. Agrawal, J. C. Parnami, D. B. Rishi, and S. A. Katre all from the Panjab School at Chandigarh, dealing with the explicit determination of the number of points on an elliptic curve with complex multiplication.

The active number theorists of the Panjab School are still involved in important research activities, but those are not discussed because of the stipulated time frame of the present project.

Finally, the major contributions of some of the number theorists of TIFR deserve special mention. K. Chandrasekharan's works with Raghavan Narasimhan were the first major research contributions in the theory of Zeta functions in India after S. Chowla. In a long series of papers, they were able to obtain for a wide class of Dirichlet series satisfying a functional equation resembling that of the Riemann zeta function, a number of key properties that were at that time known essentially only for the Riemann zeta function.

K. Ramachandra has done remarkable contributions to various areas of number theory. It is essential to mention some of his most famous works. Ramachandra made a remarkable application of Kronecker's second limit formula to the theory of complex multiplication, to the construction of a certain maximally independent set of units in a given class field of an imaginary quadratic field, and to the evaluation of a certain elliptic integral, originally given by Chowla and Selberg and also by Ramanujan. He was associated with the construction of the "Siegel–Ramachandra–Robert units". These are explicit units in abelian extensions of quadratic imaginary fields constructed using elliptic functions. Siegel did the initial work. After that, they were constructed by Ramachandra and then streamlined by Robert. For some time, they were known as "Siegel–Ramachandra–Robert units". Now they are called elliptic units and they play a fundamental role in many works dealing with the arithmetic of elliptic curves with complex multiplication. These units have been used in the fundamental works of Coates–Wiles and Rubin. Ramachandra got interested in Baker's method in transcendental number theory in the late sixties of the twentieth century. Under the guidance of Ramachandra, T. N. Shorey took up serious research in this field. Shorey is renowned for his contributions to the application of Baker's theory.

The research work carried out on Oppenheim's conjecture by the number theorists of the Panjab School of Chandigarh has already been discussed. In TIFR, S. Raghavan, K. G. Ramanathan, M. S. Raghunathan, and S. G. Dani too made remarkable contributions in this particular field. S. Raghavan and K. G. Ramanathan jointly proved an analogue over algebraic number fields of a result of A. Oppenheim. Oppenheim had earlier established a result on the density of values of indefinite quadratic forms, in $n \geq 5$ variables, which were not scalar multiples of rational forms and represented zero. Raghavan's interest in the Oppenheim conjecture, for forms that may not represent zero indirectly influenced some future developments at TIFR. The mathematicians involved in furthering work related to Oppenheim conjecture were M. S. Raghunathan and S. G. Dani. Sometime in 1975, M. S. Raghunathan had casually suggested a statement on the behavior of what are called *unipotent flows* and asked Dani to name it as *Raghunathan's conjecture* and advised him to prove it. Raghunathan also pointed out that if his conjecture could be solved, it would in particular settle the conjecture of Oppenheim on the density of values of indefinite forms at integral points. The statement of Raghunathan—the Raghunathan conjecture was first recorded by Dani in his paper titled "Invariant measures and minimal sets of horospherical flows".[33] In this paper, Dani proposed another conjecture, as a step toward proving the Raghunathan conjecture. By the 1980s, many partial results were known confirming the Oppenheim conjecture under various restrictions. But a general solution still could not be established. The research of Raghunathan and Dani influenced the pathbreaking works of G. A. Margulis and Marina Ratner in the area. G. Margulis proved the full conjecture and gave a beautiful survey of work leading to this solution in his lecture written after receiving the Fields Medal. There he explained and wrote:

> The different approaches to this and related conjectures (and theorems) involve analytic number theory, the theory of Lie groups and algebraic groups, ergodic theory, representation theory, reduction theory, geometry of numbers and some other topics.

Shortly after Margulis' breakthrough, the proof was simplified and generalized by Dani and Margulis. Marina Ratner's profound contributions in establishing the Raghunathan conjecture and its variants in the nineties of the twentieth century are considered as milestones in "homogeneous dynamics". Her work in this field has made a great impact on diverse areas of mathematics including dynamics, Diophantine approximations, Ergodic theory, geometry, Lie group theory, and so on. S. Raghavan and S. G. Dani collaborated and did commendable work on the density of orbits of irrational Euclidean frames under actions of various familiar discrete groups of significance in Diophantine approximation of systems of linear forms.

Lastly, it may be noted that in the area of Riemann zeta function, R. Balasubramanian's contributions have been highly acclaimed. Balasubramanian in collaboration with J. B. Conrey and D. R. Heath-Brown derived an asymptotic formula for the mean square of the product of the Riemann zeta function and a Dirichlet polynomial. This method led to an improvement of the lower estimate for the density of zeros of the Zeta function on the critical line. This has opened avenues for further research in this area of number theory.

It may be noted that Prof. R. Balasubramanian's first Ph.D. student, S. D. Adhikari's research work related to Ω-results and lattice points in ellipsoids have been acclaimed internationally and have made a notable impact on modern-day research in the area.

The functional analytic interpretation and generalization of Veronin's universality theorem on the Riemann hypothesis as done by Bhaskar Bagchi of the Indian Statistical Institute and published in *Mathematische Zeitschrift* in 1982 is a much acclaimed work and is still wisely quoted. E. Kowalski, a prominent Swiss mathematician, found this work to be sufficiently interesting, and he further generalized it

---

to the context of modular forms. The probabilistic aspect of Bagchi's thesis was also revived by Kowalski under the name *Bagchi's theorem*. The universality theorem as established by Bagchi has been reproduced in Heath-Brown's revised edition of the classic treatise of E. C. Titchmarsh named *The Riemann Zeta-function* in 1988.

Chapter 4
Epilogue

This chapter has two sections. In the first section, discussions have been made on the applications of some results of number theory established by Indian mathematicians in all real-life areas. The second section comprises concluding remarks.

4.1 Practical Applications

The theoretical impact of the researches carried out during the twentieth century by various number theorists of India on modern mathematics has been discussed elsewhere in the book.[1]

It is necessary to highlight the practical applications of the number theory-based and related research carried out during the same period by the Indian number theorists. After an intensive literature survey, it has been found that many results of Srinivasa Ramanujan on number theory and number theory-based disciplines have been used in various practical applications with great success. The applications have been made in such diverse fields as crystallography, statistical mechanics, communication networks, biology, and even in string theory. Later on, towards the end of the twentieth century, some research works by other Indian number theorists have also been applied to practical problems.

4.1.1 Crystallography

It is an experimental science of determining the arrangement of atoms in crystalline structures. Noted Indian crystallographer S. Ramaseshan has shown how Ramanujan's work on partitions sheds light on plastics. Plastics, as is well-known,

---

[1]Refer to Chap. 3.

P. Mukherji, *Research Schools on Number Theory in India*,
https://doi.org/10.1007/978-981-15-9620-9_4

are essentially polymers. They repeat molecular units that combine in various ways. Suppose there is one that is a million units long, another 8474 units long, another 2, 35, 819 units long, and so on. Ramanujan showed in his partition theory on how smaller units combine to form larger ones. So, this theory clearly bears on the said process.

4.1.2 Statistical Mechanics

The distribution of a film of liquid helium on a graphite plate is a physical example of the *hard hexagon* model. The Rogers–Ramanujan identities arise naturally in R. J. Baxter's solution of the above mentioned model in statistical mechanics. Mathematically, Baxter showed that the hard hexagon model was built on a particular set of infinite series. He also discovered that those series are exactly the ones that occur in the famous Rogers–Ramanujan identities. Predictions made by using these kinds of models agree closely with experimental results.

Another relatively recent development linking prime numbers and physics has been discussed in an article by H. Gopalkrishna Gadiyar and R. Padma. The paper titled "Ramanujan–Fourier series, the Weiner–Khintchine formula and the distribution of prime pairs" jointly written by them was published in *Physica A*.[2] In the abstract, the authors explain that the *Weiner–Khintchine formula* is a central piece in statistical mechanics. In the above mentioned paper, the authors have shown that the problem of prime pairs is related to autocorrelation and hence to the Weiner–Khintchine formula. They have also provided experimental evidence for the same. The authors have observed that:

> It is a pleasant surprise that Weiner–Khintchine formula which normally occurs in practical problems of Brownian motion, electrical engineering and other applied areas of technology and statistical physics has a role in the behaviour of prime numbers which are studied by pure mathematicians.

4.1.3 Computation of the Value of π (Pai)

Ramanujan's 1914 publication on *Modular Equations and Approximations to Pi* gave the algorithms for evaluation π that are fastest in use today. Very recently, Ramanujan's transformations for elliptic functions were used by David and Gregory Chudnovsky to produce very rapidly convergent algorithms to compute π. Then Chudnovskys has now computed π to the order of a billion digits.

---

[2]Vol. 269, (1999), 503–510.

4.1.4 Ramanujan Graphs

The mathematical discipline needed for the study of the complex networks in biology, communications, and elsewhere is graph theory. The knowledge of elementary number theory is mandatory for learning graph theory.

Ramanujan graphs are extensively used in communication networks. Ramanujan graphs are regular graphs with small non-trivial eigenvalues. They are regular to regular random graphs. They are used in communication networks and also in the construction of low-density parity check codes. A central problem in communication technology is the construction of efficient networks at a cost not exceeding a fixed amount. Mathematically, the networks may be represented by a graph G, and the efficiency may be measured by the magnifying constant C of the graph. The desired exercise is to have explicit construction of graphs with a lower bound of the magnifying constant. For the explicit construction of Ramanujan graphs, there are three systematic ways in use. They are constructed using number theoretic methods listed below:

(a) Ramanujan graphs based on adelic quaternionic groups,
(b) Ramanujan graphs based on finite Abelian groups,
(c) Ramanujan graphs based on finite non-Abelian groups.

Apart from these applications, of late, Ramanujan graphs have opened up new vistas for other applications. The theory of complex networks plays an important role in a wide variety of disciplines, ranging from communications and power system engineering to molecular and population biology. For instance, in recent years the Internet and the World Wide Web (www) have grown at a remarkable rate both in importance as well as volume. Again in sociology and ecology, an increasing amount of data on the food-web and the structure of human social networks are now available. For these reasons, network analysis has become a very important exercise. The threats to human health posed by new infectious diseases such as the Asian bird flu and swine flu coupled with modern travel patterns highlight the importance of these issues.

4.1.5 Lattice

Lattices are regular arrangements of atoms, ions, or molecules in a crystalline solid. A lattice model in physics refers to a model which is not defined on a continuum but on a grid. Lattices arise in many areas of number theory and physics. One reason for their presence everywhere is because the lattice is the basic framework for periodic structures, both geometric as well as algebraic. They are important because many non-periodic structures can be described as irrational sections of lattices in higher dimensions. In collaboration with K. S. Viswanathan (a former student of Sir C. V. Raman), famous crystallographer S. Ramaseshan started studying lattice dynamics and structure of non-crystalline materials. They established the breakdown of Friedel's law in elastic neutron scattering.

4.1.6   Theoretical Physics

Ramanujan's work has continued to have a significant impact on theoretical physics in the ongoing century too, especially in the area of quantum black holes in string theory. Ever since the seminal work of Stephen Hawking, it is well understood that black holes have thermodynamic properties such as entropy, which is directly proportional to the area of the horizon of the black hole. Given the prevalent understanding of thermodynamics, it is natural to wonder what microscopic states are responsible for this entropy.

For special classes of "supersymmetric",[3] black holes in string theory, the question about the origin of this degeneracy of states has been answered at the microscopic level. Furthermore, what has been shown is that the solution is intimately related to the mathematics of mock-modular forms in the following way: the generating function that accounts for the degeneracy of states for such black holes in string theory is an almost modular function that develops jumps or discontinuities as certain parameters of the black hole are varied. Rather miraculously, precisely such functions were first introduced by Ramanujan in one of his last letters to Hardy in 1920, in which he christened them mock-theta functions.

While Ramanujan gave as many as 17 examples of these mock-theta functions in his letters to Hardy, a complete definition was not given by him, and it was not until the thesis work of Sander Zwegers[4] that these were placed on a firm mathematical footing. Building upon these results, it was shown by Dabholkar, Murthy, and Zagier that quantum degeneracies of single-centered black holes in a class of supersymmetric string theories are given by Fourier coefficients of a mock-modular form.

4.2   Concluding Remarks

Apart from the remarkable contributions and steady growth of schools of research on number theory in India during the twentieth century, it may be worthwhile to mention a few things in this context.

---

[3] *Supersymmetry* is a principle that proposes a relationship between two basic classes of elementary particles: bosons that have integer spin and fermions that have half-integer spin. While there is as yet no experimental evidence for supersymmetry, it provides greater analytic control and often leads to exactly solvable physical models.

[4] Ph.D. thesis titled "Mock Theta functions", Utrecht University, 2002.

In the present century, too, the Indian number theorists are carrying on important research work that is acclaimed internationally. The practical applications of number theory have also extended to areas such as coding theory, cryptology, and cyber security-related matters. Indian mathematicians are making valuable contributions in those areas, too.

Another point needs to be definitely mentioned. It has traditionally been the case that questions in number theory have provided the motivation for the development of large tracts of mathematics, and in turn, number theory has used a lot of the developments in various branches of mathematics. This aspect has become all the more important since the 1960s with the work of Weil, Grothendieck, and Langlands. The voracious appetite of number theory for mathematics has led to a situation where much of modern mathematics can now be considered as part of number theory. To illustrate the work of lattices, Lie groups are foundational for the study of automorphic forms; the work on vector bundles has been instrumental in proving the fundamental lemma by Ngo Bao Chau; the work on ergodic theory in the context of homogeneous spaces has had innumerable arithmetical applications, and so on.

Bibliography

Section 1

Srinivasa Ramanujan (SR)

1. "Some properties of Bernoulli's numbers". *Journal of the Indian Mathematical Society*. 3, (1911), 219–234.
2. "Irregular numbers". *Journal of the Indian Mathematical Society*. 5, (1913), 105–106.
3. "Squaring the circle". *Journal of the Indian Mathematical Society*. 5, (1913), 132.
4. "Modular equations and approximations to π". *Quarterly Journal of Mathematics*. 45, (1914), 350–372.
5. "On the number of divisors of a number". *Journal of the Indian Mathematical Society*. 7, (1915), 131–133.
6. "On the sum of the square roots of the first n natural numbers". *Journal of the Indian Mathematical Society*. 7, (1915), 173–175.
7. "New expressions for Riemann's functions $\xi(s)$ and $\Xi(t)$". *Quarterly Journal of Mathematics*. 46, (1915), 253–260.
8. "Highly composite numbers". *Proceedings of the London Mathematical Society*, 2. 14, (1915), 347–409.
9. "Some formulae in the analytic theory of numbers". *Messenger of Mathematics*. 45, (1916), 81–84.
10. "On certain arithmetical functions". *Transactions of the Cambridge Philosophical Society*. Vol. 22, No. 9, (1916), 159–184.
11. "On the expression of a number in the form $ax^2 + by^2 + cz^2 + du^2$". *Proceedings of the Cambridge Philosophical Society*. 19, (1917), 11–21.
12. (with G. H. Hardy) "Une formulae asymptotique pour le nombre des partitions de n". *Comptes Rendus*. 2 January (1917).
13. (with G. H. Hardy) "Proof that almost all numbers n are composed of about log log n prime factors". *Proceedings of the London Mathematical Society*, 2. 16, (1917).
14. (with G. H. Hardy) "Asymptotic formulae in combinatory analysis". *Proceedings of the London Mathematical Society*, 2. 16, (1917).

15. (with G. H. Hardy) "Asymptotic formulae for the distribution of integers of various types". *Proceedings of the London Mathematical Society*, 2. 16, (1917), 112–132.
16. (with G. H. Hardy) "The normal number of prime factors of a number n". *Quarterly Journal of Mathematics*. 48, (1917), 76–92.
17. (with G. H. Hardy) "Asymptotic formulae in combinatory analysis". *Proceedings of the London Mathematical Society*, 2. 17, (1918), 75–115.
18. (with G. H. Hardy) "On the coefficients in the expansions of certain modular functions". *Proceedings of the Royal Society*, A. 95, (1918), 144–155.
19. "On certain trigonometrical sums and their applications in the theory of numbers". *Transactions of the Cambridge Philosophical Society*, 22. 13, (1918), 259–276.
20. "Some properties of $p(n)$, the number of partitions of n". *Proceedings of the Cambridge Philosophical Society*. 19, (1919), 207–210.
21. "Proof of certain identities in combinatory analysis". *Proceedings of the Cambridge Philosophical Society*. 19, (1919), 214–216.
22. "Congruence properties of partitions". *Proceedings of the London Mathematical Society*, 2. 18, (1920).
23. "Congruence properties of partitions". *Mathematische Zeitschrift*. (1921), 147–153.

K. Ananda Rau (KAR)

1. "The infinite product for $(s - 1)\zeta(s)$". *Mathematische Zeitschrift*. 20, (1924), 156–164.
2. "On the boundary behaviour of elliptic modular functions". *Acta Mathematica*. 52, (1929), 143–168.
3. "Additional note on the boundary behaviour of elliptic modular functions". *Acta Mathematica*. 53, (1929), 77–86.
4. "On the behaviour of elliptic theta functions near the line of singularities". *Journal of the Indian Mathematical Society*, 20, (1933), 148–156.
5. "On the representation of a number as the sum of an even number of squares". *Journal of Madras University*. B, 24, (1954), 61–89.
6. "On the summation of singular series associated with certain quadratic forms I". *Journal of the Indian Mathematical Society*. (N. S.) 23, (1959), 65–96.
7. "Application of modular equations to some quadratic forms". *Journal of the Indian Mathematical Society*. (N. S.) Jubilee Issue, 24, (1960), 77–130.
8. "Relation between sums of singular series associated with certain quadratic forms". *Journal of Madras University*. B, 31, (1961), 7–10.
9. "On the summation of singular series associated with certain quadratic forms II". *Journal of the Indian Mathematical society*. (N. S.), 25, (1961), 173–195.

T. Vijayaraghavan (TV)

1. "Periodic simple continued fractions". *Proceedings of the London Mathematical Society*. 26, (1927), 403–414.
2. "A note on diophantine approximation". *Journal of the London Mathematical Society*. 2, (1927), 13–17.
3. "On the irrationality of a certain decimal", *Proceedings of the Indian Academy of Sciences*, A. 10, (1939), 341.
4. "On decimals of irrational numbers". *Proceedings of the Indian Academy of Sciences*, A. 12, (1940), 20.
5. "On the fractional parts of the powers of a number, I". *Journal of the London Mathematical Society*. 15, (1940), 159–160.
6. "On Jaina magic squares". *Mathematics Student*. 9, (1941), 97–102.
7. "On the fractional parts of the powers of a number, II". *Proceedings of the Cambridge Philosophical Society*. 37, (1941), 349–357.
8. "On the fractional parts of the powers of a number III". *Journal of the London Mathematical Society*. 17, (1942), 137–138.

9. (with S. Chowla) "The complex factorization (mod p) of the cyclotomic polynomial of order $p^2 - 1$". *Proceedings of the National Academy of Sciences, India,* A. 14, (1944), 101–105.
10. (with S. Chowla) "Short proofs of theorems of Bose and Singer". *Proceedings of the National Academy of Sciences, India,* A. 15, (1945), 194.
11. "On the largest prime divisors of numbers". *Journal of the Indian Mathematical Society.* 11, (1947), 31–37.
12. "On the fractional parts of the powers of a number, IV". *Journal of the Indian Mathematical Society.* 12, (1948), 33–39.
13. (with S. Chowla) "On complete residue sets". *Quarterly Journal of Mathematics.* Oxford, 19, (1948), 193–199.
14. "On a problem in elementary number theory". *Journal of the Indian Mathematical Society.* 15, (1951), 51–56.

S. S. Pillai (SSP)

1. "A test for groups of primes". *Journal of the Indian Mathematical Society.* 17, (1927–28), 85–88.
2. "On some empirical theorems of Scherk". *Journal of the Indian Mathematical Society.* 17, (1927–28), 164–171.
3. "On the representation of a number as the sum of two positive k-th powers". *Journal of the London Mathematical Society.* 3, (1928), 56–61.
4. "Corrigenda:on the representaion of a number as the sum of two positive k-th powers". *Journal of the London Mathematical Society.* 3, (1928), 83.
5. "On some functions connected with $\varphi(n)$". *Bulletin of the American Mathematical Society.* 35, (1929), 832–836.
6. "On a function connected with $\varphi(n)$". *Bulletin of the American Mathematical Society.* 35, (1929), 837–841.
7. "On the number of numbers which contain a fixed number of prime factors". *Mathematics Student.* 14, (1929), 250–251.
8. "A theorem concerning the primitive periods of integer matrices". *Journal of the London Mathematical Society.* 4, (1929), 250–251.
9. (with S. Chowla) "On the error terms in some asymptotic formulae in the theory of numbers I". *Journal of the London Mathematical Society.* 5, (1930), 95–101.
10. "On the numbers which contain no prime factors of the form $p(kp + 1)$". *Journal of the Indian Mathematical Society,* 18, (1930), 51–59.
11. (with S. Chowla)"On the error terms in some asymptotic formulae in the theory of numbers II". *Journal of the Indian Mathematical Society.* 18, (1930), 181–184.
12. "On a function analogues to $G(k)$". *Journal of the Indian Mathematical Society.* 18, (1930), 289–290.
13. "On some diophantine equations". *Journal of the Indian Mathematical Society.* 18, (1930), 291–295.
14. "On the inequality $0 < a^x - b^y < n$". *Journal of the Indian Mathematical Society.* 19, (1931), 1–11.
15. (with S. Chowla) "Periodic simple continued fractions". *Journal of the London Mathematical Society.* 6, (1931), 85–89.
16. "An order-result concerning φ-function". *Journal of the Indian Mathematical Society.* 19, (1931), 165–168.
17. "On the sum function of the number of prime factors of N". *Journal of the Indian Mathematical Society.* 20, (1932), 70–87.
18. "On the indeterminate equation $x^y - y^x = a$". *Journal, Annamalai University, I.* (1), (1932), 59–61.
19. "On an arithmetic function concerning primes". *Journal, Annamalai University, I.* (2), (1932), 159–167.
20. "On an arithmetic function". *Journal, Annamalai University, II.* (2), 242–248.

21. "Periodic simple continued fractions". *Journal, Annamalai University, IV*. (2), (1935), 216–225.
22. "On Waring's problem I". *Journal, Annamalai University, V*. (2), (1936), 145–166.
23. "On Waring's problem II". *Journal of the Indian Mathematical Society* (N. S.), II (1936), 16–44.
24. "On Waring's problem III". *Journal, Annamalai University, VI*. (1), (1936), 50–53.
25. "On Waring's problem IV". *Journal, Annamalai University, VI*. (1), (1936), 54–64.
26. "On the set of square-free numbers". *Journal of the Indian Mathematical Society* (N. S.), II (1936), 116–118.
27. "On Waring's problem V: on $g(6)$". *Journal of the Indian Mathematical Society* (N. S.), II. (1936), 213–214.
28. 'On $a^x - b^y = c$". *Journal of the Indian Mathematical Society* (N. S.), II. (1936), 119–122.
29. "Correction to the paper 'On$a^x - b^y = $ c'". *Journal of the Indian Mathematical Society* (N. S.), II. (1936), 215.
30. (with S. Chowla) "Hypothesis K of Hardy and Littlewood". *Mathematische Zeitschrift*, 41, (1936), 537–540.
31. (with S. Chowla) "The number of representations of a number as a sum of n non-negative n-th powers". *Quarterly Journal of Mathematics*, Oxford, Ser. 7, (1936), 56–59.
32. "On Waring's problem VI". *Journal, Annamalai University, VII*. (1937), 171–197.
33. "Generalization of a theorem of Davenport on the addition of residue classes". *Proceedings of the Indian Academy of Sciences, Sect. A*. VI, (1937), 179–180.
34. "On the addition of residue classes". *Proceedings of the Indian Academy of Sciences, Sect. A*. VII, (1938), 1–4.
35. "On Waring's problem with powers of primes I". *Proceedings of the Indian Academy of Sciences, Sect. A*. IX, (1939), No. 1, 29–34.
36. "On $v(k)$". *Proceedings of the Indian Academy of Sciences, Sect. A*. IX, (1939), 175–176.
37. "On normal numbers I". *Proceedings of the Indian Academy of Sciences, Sect. A*. X, (1939), 13–15.
38. "On the smallest prime of the form $km + 1$". *Proceedings of the Indian Academy of Sciences, Sect. A*. X, (1939), 388–389.
39. "A note on the paper of Sambasiva Rao". *Journal of the Indian Mathematical Society* (N. S.), III, (1939), 266–267.
40. "On the number of representations of a number as the sum of the square of a prime and square-free integer". *Proceedings of the Indian Academy of Sciences, Sect. A*. X, (1939), 390–391.
41. "On numbers which are not multiples of any other in the set". *Proceedings of the Indian Academy of Sciences, Sect. A*. X, (1939), 392–394.
42. "On Waring's problem VIII (with polynomial summonds)". *Journal of the Indian Mathematical Society* (N. S.), III, (1939), 205–220. [In the sequel, "On Waring's problem VII" is missing.]
43. "On Waring's problem IX (on universal Waring's problem with prime powers)". *Journal of the Indian Mathematical Society* (N. S.), III, (1939), 221–225.
44. "On Stirling's approximation". *Mathematics Student*. VII, (1939), 70–71.
45. "Symposium on Waring's problem—P. Chairman's address". *Mathematics Student*. VII, (1939), 165–168.
46. "On the converse of Fermat's theorem". *Mathematics Student*. VIII, (1940), 132–133.
47. "On $m$consecutive integers I". *Proceedings of the Indian Academy of Sciences, Sect. A*. XI, (1940), 6–12.
48. "Generalisation of a theorem of Mangoldt". *Proceedings of the Indian Academy of Sciences, Sect. A*. XI, (1940), 13–20.
49. "On $m$consecutive integers II". *Proceedings of the Indian Academy of Sciences, Sect. A*. XI, (1940), 73–80.
50. "On Waring's problem $g(6) = 73$". *Proceedings of the Indian Academy of Sciences, Sect. A*. XII, (1940), 30–40.

51. "Waring's problem with indices $\geq n$". *Proceedings of the Indian Academy of Sciences, Sect. A.* XII, (1940), 41–45.
52. "A note on Gupta's previous paper". *Proceedings of the Indian Academy of Sciences, Sect. A.* XII, (1940), 63–64.
53. "On normal numbers II". *Proceedings of the Indian Academy of Sciences, Sect. A.* XII, (1940), 179–184.
54. "On a linear diophantine equation". *Proceedings of the Indian Academy of Sciences, Sect. A.* XII, (1940), 199–201.
55. "On Waring's problem with powers of primes II". *Proceedings of the Indian Academy of Sciences, Sect. A.* XII, (1940), 202–204.
56. "On the sum function connected with primitive roots". *Proceedings of the Indian Academy of Sciences, Sect. A.* XIII, (1941), 526–529.
57. "On m consecutive integers III". *Proceedings of the Indian Academy of Sciences, Sect. A.* XIII, (1941), 530–533.
58. "On the definition of oscillation". *Mathematics Student.* IX, (1941), 165–167.
59. "On numbers of the form $2^a 3^b$–I". *Proceedings of the Indian Academy of Sciences, Sect. A.* XV, (1942), 128–132.
60. (with George, A.) "On numbers of the form $2^a 3^b$–II". *Proceedings of the Indian Academy of Sciences, Sect. A.* XV, (1942), 133–134.
61. "On algebraic irrationals". *Proceedings of the Indian Academy of Sciences, Sect. A,* XV, (1942), 173–176.
62. "On a problem in diophantine approximation". *Proceedings of the Indian Academy of Sciences, Sect. A,* XV, (1942), 177–189.
63. "On a congruence property of a divisor function". *Journal of the Indian Mathematical Society* (N. S.), VI, (1942), 118–119.
64. "On the divisors of $a^n + 1$". *Journal of the Indian Mathematical Society* (N. S.), VI, (1942), 120–121.
65. "*Lattice points in a right-angled triangle II*". *Proceedings of the Indian Academy of Sciences, Sect. A.* XVII, (1943), 58–61.
66. "Lattice points in a right-angled triangle III". *Proceedings of the Indian Academy of Sciences, Sect. A.* XVII, (1943), 62–65.
67. "On $\sigma_{-1}(n)$ and $\varphi(n)$". *Proceedings of the Indian Academy of Sciences, Sect. A.* XVII, (1943), 67–70.
68. "Highly abundant numbers". *Bulletin of the Calcutta Mathematical Society.* 35, (1943), 141–156.
69. "On the smallest primitive root of a prime". *Journal of the Indian Mathematical Society* (N. S.). VIII, (1944), 14–17.
70. "On Waring's problem with powers of primes III". *Journal of the Indian Mathematical Society* (N. S.). VIII, (1944), 18–20.
71. "Highly composite numbers of the t-th order". *Journal of the Indian Mathematical Society* (N. S.). VIII, (1944), 61–74.
72. "Bertrand's postulate". *Bulletin of the Calcutta Mathematical Society.* 37, (1944), 97–99.
73. "On m consecutive integers IV". *Bulletin of the Calcutta Mathematical Society.* 37, (1944), 99–101.
74. "On the equation $2^x - 3^y = 2^X + 3^Y$". *Bulletin of the Calcutta Mathematical Society.* 37, (1945), 15–20.

Section 2

S. Chowla (SC)

[Solutions by S. Chowla to problems posed in the *Journal of the Indian Mathematical Society* (1925–1931)].

S1.   (With S. Audinaraniah, T. Totadri Aiyengar) "Solution of question no. 1252 (Sanjana)". *Journal of the Indian Mathematical Society (Notes and Questions)*. 16, (1925), 54.

S2.   "Solution of question no. 1298 (T. Iyengar)". *Journal of the IndianMathematical Society (Notes and Questions)*. 16, (1925), 76.

S3.   "Solution of question no. 1344 (Tiruvenkatacharya)". *Journal of the Indian Mathematical Society (Notes and Questions)*. 16, (1925), 89.

S4.   "Solution of question no. 1353 (Enquirer)". *Journal of the Indian Mathematical Society (Notes and Questions)*. 16, (1925), 90–92.

S5.   "Remarks on question no. 353 (Ramanujan)". *Journal of the Indian Mathematical Society (Notes and Questions)*. 16, (1926), 119–120.

S6.   (With N. B. Mitra, S. V. Venkataraya Sastri) "Solution of question no. 1070 (Ramaujan)". *Journal of the Indian Mathematical Society (Notes and Questions)*. 16, (1926), 122–123.

S7.   (With V. Tiruvenkatachariar) "Solutions of questions nos. 1084, 1085 and 1086 (Hemraj)". *Journal of the Indian Mathematical Society (Notes and Questions)*. 16, (1926), 155–157.

S8.   (With R. Srinivasav, H. R. Gupta, G. V. Krishnaswamy, M. V. Seshadri,, V. A. Mahalingam, P. Kameswara Rao) "Solutions of question no. 1331 (Satyanarayana)". *Journal of the Indian Mathematical Society (Notes and Questions)*. 17, (1927), 14–15.

S9.   "Solution of question No. 1367 (Chowla)". *Journal of the Indian Mathematical Society (Notes and Questions)*. 17, (1927), 15.

S10. (With P. Kameswara Rao, V. A. Mahalingam) "Solution of question no. 1332 (Satyanarayana)". *Journal of the Indian Mathematical Society (Notes and Questions)*. 17, (1927), 46.

S11. "Solution of question no. 1007 (Trivedi)". *Journal of the Indian Mathematical Society (Notes and Questions)*. 17, (1927), 58–59.

S12. "Solution of question no. 1385 (Thiruvenkatacharya)". *Journal of the Indian Mathematical Society (Notes and Questions)*. 17, (1927), 93–94.

S13. (With T. R. Raghavasastri, T. Totadri Aiyengar, T. Vijayaraghavan) "Solution of question no. 1437 (Ananda Rau)". *Journal of the Indian Mathematical Society (Notes and Questions)*. 17, (1928), 109.

S14. "Remarks on question no. 629 (Ramanujan)". *Journal of the Indian Mathematical Society (Notes and Questions)*. 17, (1928), 136–137.

S15. (With S. Mahadevan, Kanwar Bahadur, T. Totadri Aiyengar) "Solution of question no. 1415 (Audinarayanan)". *Journal of the Indian Mathematical Society (Notes and Questions)*. 17, (1928), 158.

S16. "Solution and remarks on question no. 770 (Ramanujan)". *Journal of the Indian Mathematical Society (Notes and Questions)*. 17, (1928), 166–171.

S17. "Solution of question no. 1255 (Vijayaraghavan)". *Journal of the Indian Mathematical Society (Notes and Questions)*. 17, (1928), 186.

S18. "Remarks on question no. 1280 (Bheemasena Rao and Krishnamachari)". *Journal of the Indian Mathematical Society (Notes and Questions)*. 17, (1928), 187.

S19. (With S. Mahadevan) "Solution of question no. 1303 (Thiruvenkatacharya)". *Journal of the Indian Mathematical Society (Notes and Questions)*. 17, (1928), 188.

S20. (With M. V. Seshadri, S. Mahadevan, Kanwar Bahadur) "Solution of question no. 1391 (Thiruvenkatacharya)". *Journal of the Indian Mathematical Society (Notes and Questions)*. 17, (1928), 190.

S21. "Solution of question no. 1019 (Hemraj)". *Journal of the Indian Mathematical Society (Notes and Questions)*. 18, (1929), 41–42.

S22. "Solution of question no. 1143 (Ananda Rao)". *Journal of the Indian Mathematical Society (Notes and Questions)*. 18, (1929), 42–43.

S23. "Remarks on question no. 1348 (Chowla)". *Journal of the Indian Mathematical Society (Notes and Questions)*. 18, (1929), 44.

S24. "Solution of question no. 1390 (Chowla)". *Journal of the Indian Mathematical Society (Notes and Questions)*. 18, (1929), 61.

S25. "Solution and remarks on question no. 1402 (Chowla)". *Journal of the Indian Mathematical Society (Notes and Questions)*. 18, (1929), 63–65.

S26. "Remarks on question no. 1444". *Journal of the Indian Mathematical Society (Notes and Questions)*. 18, (1929), 94.

S27. "Solution of question no. 1489 (Pillai)". *Journal of the Indian Mathematical Society (Notes and Questions)*. 18, (1930), 222.

S28. "Remarks on question no. 1490 (Pillai)". *Journal of the Indian Mathematical Society (Notes and Questions)* 18, (1930), 223.

S29. "Solution of question no. 1488 (Malurkar)". *Journal of the Indian Mathematical Society (Notes and Questions)* 18, (1930).

S30. "Solution of question no. 1527 (Vaidyanathaswamy)". *Journal of the Indian Mathematical Society (Notes and Questions)*. 19, (1931), 52.

S31. (With P. Jagannathan) "Solution of question no. 1563 (Vaidyanathaswamy)". *Journal of the Indian Mathematical Society (Notes and Questions)*. 19, (1931), 54–55.

S32. (With S. Mahadevan) "Solution of question no. 1487 (Chowla)". *Journal of the Indian Mathematical Society (Notes and Questions)*. 19, (1931), 72–75.

S33. (With Budharam, Hukam Chand, K. V. Vedantham, N. P. Subramaniam, H. R. Gupta) "Solution of question no. 1500 (Narayana Aiyar)". *Journal of the Indian Mathematical Society (Notes and Questions)*. 19, (1931), 77.

(ii) Mathematical Papers and Notes (1926–1949).

[Professor Chowla left India after the partition of the country. He settled down permanently in the USA. So, his publications till 1949 only have been listed.]

1. "Some results involving prime numbers". *Journal of the Indian Mathematical Society (Notes and Questions)*. 16, (1926), 100–104.

2. "A new proof of Von Staudt's theorem". *Journal of the Indian Mathematical Society (Notes and Questions)*. 16, (1926), 145–146.

3. "Gauss's formula and allied results". *Journal of the Indian Mathematical Society (Notes and Questions)*. 17, (1927), 4–7.

4. "An elementary treatment of the modular equation of the third order". *Journal of the Indian Mathematical Society (Notes and Questions)*. 17, (1927), 37–40.

5. "On the order of $d(n)$, the number of divisors of n". *Journal of the Indian Mathematical Society (Notes and Questions)*. 17, (1927), 55–57.

6. "Some properties of Eulerian and prepared Bernoullian numbers". *Messenger of Mathematics*. 57, (1927), 121–126.

7. "Some applications of the Riemann zeta and allied functions". *Tohoku Mathematical Journal*. 30, (1928), 202–225.

8. "On some identities involving zeta-functions". *Journal of the Indian Mathematical Society*. 17, (1928), 153–163.

9. "On a formula due to T. Ono". *Journal of the Indian Mathematical Society (Notes and Questions)*. 17, (1928), 113–115.

10. "On a certain limit connected with pairs of integers". *Journal of the Indian Mathematical Society*. 18, (1929), 13–15.

11. "Some properties of Eulerian numbers". *Tohoku Mathematical Journal*. 30, (1929), 324–327.

12. "Some identities in the theory of numbers". *Journal of the Indian Mathematical Society (Notes and Questions)*. 18, (1929), 87–88.

13. "On the greatest prime factor of a certain product". *Journal of the Indian Mathematical Society*. 18, (1929), 135–137.

14. "An order result involving Euler's φ-function'. *Journal of the Indian Mathematical Society*. 18, (1929), 138–141.

15. "On the order of $N(R)$, the number of terms in the period of the continued fraction for $\sqrt{R}$". *Journal of the Indian Mathematical Society*. 18, (1929), 142–144.

16. "Expressions for the class number of binary quadratic forms". *Journal of the Indian Mathematical Society*. 18, (1929), 145–146.
17. "An elementary note on Waring's theorem on cubes". *Journal of the Indian Mathematical Society (Notes and Questions)*. 18, (1929), 126–128.
18. "A property of positive odd integer". *Journal of the Indian Mathematical Society (Notes and Questions)*. 18, (1929), 129–130.
19. (With S. S. Pillai) "On the error terms in some asymptotic formulae in the theory of numbers (I)". *Journal of the London Mathematical Society*. 5, (1930), 95–101.
20. (With S. S. Pillai)"On the error terms in some asymptotic formulae in the theory of numbers (II)". *Journal of the Indian Mathematical Society*. 18, (1930), 181–184.
21. "A congruence theorem". *Journal of the Indian Mathematical Society (Notes and Questions)*. 18, (1930), 145–146.
22. "Remarks on Waring's theorem". *Journal of the London Mathematical Society*. 5, (1930), 155–158.
23. "A generalization of a theorem of Wolstenholme". *Journal of the London Mathematical Society*. 5, (1930), 158–160.
24. "Cauchy's criterion for the solvability of $x^p + y^p = z^p$ in integers prime to p". *Journal of the Indian Mathematical Society*. 18, (1930), 205–206.
25. "On a conjecture of Ramanujan". *Tohoku Mathematical Journal*. 32, (1930), 1–2.
26. "Some problems of Diophantine approximation (I)". *Mathematische Zeitschrift*. 33, (1931), 544–563.
27. (With S. S. Pillai) "Periodic simple continued fractions". *Journal of the London Mathematical Society*. 6, (1931), 85–89.
28. "Two problems in the theory of lattice points". *Journal of the Indian Mathematical Society*. 19, (1931), 97–108.
29. "Contributions to the analytic theory of numbers". *Mathematische* Zeitschrift. 35, (1932), 279–299.
30. "A theorem on characters (II)". *Journal of the Indian Mathematical Society*. 19, (1932), 279–284.
31. "Contributions to the analytic theory of numbers (II)". *Journal of the Indian Mathematical Society*. 20, (1933), 121–128.
32. "Notes on the theory of numbers (I): prime numbers". *Mathematics Student*. 1, (1933), 41–48.
33. (With A. Sreerama Sastri) "A proof of the theorem of Wolstenholme". *Mathematics Student*. 1, (1933), 106–107.
34. "A generalization of a theorem of Wolstenholme". *Mathematics Student*. 1, (1933), 140–141.
35. "Primes in arithmetical progression". *Mathematics Student*. 1, (1933), 147.
36. "On the k-analogue of a result in the theory of Riemann zeta-function". *Mathematische Zeitschrift*. 38, (1934), 483–487.
37. "A theorem on characters". *Tohoku Mathematical Journal*. 39, (1934), 248–252.
38. "On the least prime in an arithmetical progression". *Journal of the Indian Mathematical Society* (N. S.). 1, (1934), 1–3.
39. "On abundant numbers". *Journal of the Indian Mathematical Society* (N. S.). 1, (1934), 41–44.
40. "A theorem on irrational indefinite quadratic forms". *Journal of the London Mathematical Society*. 9, (1934), 162–163.
41. "A theorem in arithmetic". *Journal of the London Mathematical Society*. 9, (1934), 163.
42. "The rational solution of $ax^n - by^n = k$". *Indian Physico-Mathematical Journal*. 5, (1934), 5–6.
43. "Congruence properties of partitions". *Mathematics Student*. 2, (1934), 22.
44. "Notes on the theory of numbers (II): remarks on the preceding paper". *Mathematics Student*. 2, (1934), 23.
45. "The class-number of binary quadratic forms". *Quarterly Journal of Mathematics*. Oxford, Ser 5, (1934), 302–303.

46. "Leudesdorf's generalization of Wolstenholme's theorem". *Journal of the London Mathematical Society.* 9, (1934), 246.

47. "Congruence properties of partitions". *Journal of the London Mathematical Society.* 9, (1934), 247.

48. "Some theorems in the analytic theory of numbers". *Indian Physico-Mathematical Journal.* 5, (1934), 7–8.

49. "Primes in an arithmetical progression". *Indian Physico-Mathematical Journal.* 5, (1934), 35–43.

50. "Heilbronn's class-number theorem". *Proceedings of the Indian Academy of Sciences, Sect. A.* 1, (1934), 74–76.

51. "An extension of Heilbronn's class-number theorem". *Mathematics Student.* 2, (1934), 66.

52. "Heilbronn's class-number theorem". *Journal of the Indian Mathematical Society* (N. S.). 1, (1934), 66–68.

53. "An extension of Heilbronn's class-number theorem". *Journal of the Indian Mathematical Society* (N. S.). 1, (1934), 88–92.

54. "An extension of Heilbronn's class-number theorem". *Proceedings of the Indian Academy of Sciences, Sect. A,* 1, (1934), 143–144.

55. "Heilbronn's class-number theorem (II)". *Proceedings of the Indian Academy of Sciences, Sect. A.* 1, (1934), 145–146.

56. "An extension of Heilbronn's class-number theorem". *Quarterly Journal of Mathematics.* Oxford, Ser 5, (1934), 304–307.

57. "An extension of Heilbronn's class-number theorem". *Indian Physico-Mathematical Journal.* 5, (1934), 53–57.

58. "The greatest prime factor of $x^2 - 1$". *Proceedings of the Indian Academy of Sciences, Sect. A.* 1, (1934), 269–270.

59. "The greatest prime factor of $x^2 + 1$". *Proceedings of the Indian Academy of Sciences, Sect. A.* 1, (1934), 271–273.

60. "The class-number theory of binary quadratic forms". *Proceedings of the Indian Academy of Sciences, Sect. A.* 1, (1934), 387–389.

61. (With A. Walfisz) "über eine Riemannsche Identität". *Acta Arithmetica.* 1, (1935), 87–112.

62. "Note on Dirichlet's *L*-functions". *Acta Arithmetica.* 1, (1935), 113–114.

63. "The representation of a number as a sum of four squares and a prime". *Acta Arithmetica.* 1, (1935), 115–122.

64. "The greatest prime factor of $x^2 + 1$". *Journal of the London Mathematical Society.* 10, (1935), 117–120.

65. "The representation of a large number as a sum of 'almost equal' cubes". *Quarterly Journal of Mathematics.* Oxford, Ser 6, (1935), 146–148.

66. (With S. Sastry) "On sums of powers". *Proceedings of the Indian Academy of Sciences, Sect. A.* 1, (1935), 534–535.

67. "The lattice points in a hypersphere". *Proceedings of the Indian Academy of Sciences, Sect. A.* 1, (1935), 562–566.

68. "Proof that every large number is the sum of eight 'almost equal' cubes". *Indian Physico-Mathematical Journal.* 6, (1935), 1–2.

69. (With S. Sastry) "Note on hypothesis K of Hardy and Littlewood". *Indian Physico-Mathematical Journal.* 6, (1935), 3.

70. (With S. Sastry) "Note on hypothesis K of Hardy and Littlewood". *Mathematische Zeitschrift.* 40, (1935), 348.

71. "On sums of powers (II)". *Proceedings of the Indian Academy of Sciences, Sect. A.* 1, (1935), 590–591.

72. "Note on hypothesis K of Hardy and Littlewood". *Proceedings of the Indian Academy of Sciences, Sect. A.* 1, (1935), 592.

73. "An easier Waring's problem". *Indian Physico-Mathematical Journal.* 6, (1935), 5–7.

74. "A theorem on sums of powers with applications to the additive theory of numbers". *Proceedings of the Indian Academy of Sciences, Sect. A.* 1, (1935), 698–700.

75. "A theorem on sums of powers with applications to the additive theory of numbers (II)". *Proceedings of the Indian Academy of Sciences, Sect. A.* 1, (1935), 701–706.

76. "On a certain arithmetical function". *Proceedings of the Indian Academy of Sciences, Sect. A.* 1, (1935), 772–774.

77. "Note on Euler's conjecture". *Mathematics Student.* 3, (1935), 72.

78. "A theorem on sums of powers with applications to the additive theory of numbers (III)". *Proceedings of the Indian Academy of Sciences, Sect. A.* 1, (1935), 930.

79. "Irrational indefinite quadratic forms". *Proceedings of the Indian Academy of Sciences, Sect. A.* 2, (1935), 176–177.

80. "A remarkable property of the 'singular series' in Waring's problem and its relation to hypothesis K of Hardy and Littlewood". *Proceedings of the Indian Academy of Sciences, Sect. A.* 2, (1935), 397–401.

81. "The number of representations of a large number as a sum of n non-negative n-th powers". *Indian Physico-Mathematical Journal.* 6, (1935), 65–68.

82. (With S. S. Pillai) "Hypothesis K of Hardy and Littlewood". *Mathematische Zeitschrift.* 41, (1936), 537–540.

83. (With S. S. Pillai) "The number of representations of a number as a sum of n non-negative n-th powers". *Quarterly Journal of Mathematics.* Oxford, Ser. 7, (1936), 56–59.

84. "Pillai's exact formula for the number $g(n)$ in Waring's problem". *Proceedings of the Indian Academy of Sciences, Sect. A.* 3, (1936), 339–340.

85. "Note on Waring's problem". *Proceedings of the Indian Academy of Sciences, Sect. A.* 4, (1936), 173.

86. "Pillai's exact formula for the number $g(n)$ in Waring's problem". *Proceedings of the Indian Academy of Sciences, Sect. A.* 4, (1936), 261.

87. "On a relation between two conjectures of the theory of numbers". *Proceedings of the Indian Academy of Sciences, Sect. A.* 4, (1936), 652–653.

88. 'A theorem of Erdos". *Proceedings of the Indian Academy of Sciences, Sect. A.* 5, (1937), 37–39.

89. (With F. C. Auluck) "A property of numbers". *Proceedings of the Indian Academy of Sciences, Sect. A.* 5, (1937), 510.

90. "On some arithmetical series involving arithmetical functions". *Proceedings of the Indian Academy of Sciences, Sect. A.* 5, 511–513.

91. "On some arithmetical series involving arithmetical functions (II)". *Proceedings of the Indian Academy of Sciences, Sect. A.* 5, 514–516.

92. "Auluck's generalization of the Simson line property". *Proceedings of the Indian Academy of Sciences, Sect. A.* 6, (1937), 79–80.

93. (With F. C. Auluck) "The representation of a large number as a sum of 'almost equal'numbers". *Proceedings of the Indian Academy of Sciences, Sect. A.* 6, (1937), 81–82.

94. "A remark on $g(n)$". *Proceedings of the Indian Academy of Sciences, Sect. A.* 8, (1938), 237.

95. "A remark on $g(n)$". *Proceedings of the Indian Academy of Sciences, Sect. A.* 9, (1939), 20–21.

96. (With F. C. Auluck) "An approximation connected with exp x". *Mathematics Student.* 8, (1940), 75–77.

97. (With F. C. Auluck) "On Weierstrass approximation theorem". *Mathematics Student.* 8, (1940), 78–79.

98. (With F. C. Auluck) "Some properties of a function considered by Ramanujan". *Journal of the Indian Mathematical Society* (N. S.). 4, (1940), 169–173.

99. (With F. C. Auluck and H. Gupta) "On the maximum value of the number of partitions of n into k parts". *Journal of the Indian Mathematical Society* (N. S.). 6, (1942), 105–112.

100. (With Y. Bhalotra) "Some theorems concerning quintics insoluble by radicals". *Mathematics Student.* 10, (1942), 110–112.

101. "On the k-analogue of a result in the theory of the Riemann zeta-function". *Proceedings of the Benares Mathematical Society* (N. S.). 5, (1943), 23–27.

102. "On the k-analogue of a result in the theory of the Riemann zeta-function". *Proceedings of the Lahore Philosophical Society.* 6, (1944), no. 1, 9–12.

103. "A new case of a 'complete *l-m-n* configuration'". *Proceedings of the Lahore Philosophical Society.* 6, (1944), no. 1, 13.

104. "Another case of a 'complete *l-m-n* configuration'". *Proceedings of the Lahore Philosophical Society.* 6, (1944), no. 1, 14.

105. "Solution of a problem of Erdös and Turan in additive-number theory". *Proceedings of the National Academy of Sciences, India, Sect. A.* 14, (1944), 1–2.

106. (With A. M. Mian) "On the B_2 sequences of Sidon". *Proceedings of the National Academy of Sciences, India, Sect. A.* 14, (1944), 3–4.

107. (With A. M. Mian) "On the differential equations satisfied by certain functions". *Journal of the Indian Mathematical Society* (N. S.), 8, (1944), 27–28.

108. (With A. M. Mian) "On the differential equations satisfied by certain functions". *Proceedings of the Lahore Philosophical Society.* 6, (1944), no. 2, 9–10.

109. "The cubic character of 2 (mod *p*)". *Proceedings of the Lahore Philosophical Society.* 6, (1944), no. 2, 12.

110. "Solution of a problem of Erdös and Turan in additive-number theory". *Proceedings of the Lahore Philosophical Society.* 6, (1944), no. 2, 13–14.

111. "There exists an infinity of *S*-combinations of primes in A. P.". *Proceedings of the Lahore Philosophical Society.* 6, (1944), no. 2, 15–16.

112. "On $g(k)$ in Waring's problem". *Proceedings of the Lahore Philosophical Society.* 6, (1944), no. 2, 16–17.

113. "A property of biquadratic residues". *Proceedings of the National Academy of Sciences, India, Sect. A.* 14, (1944), 45–46.

114. (With T. Vijayaraghavan) "The complete factorization (mod *p*) of the cyclotomic polynomial of order $p^2 - 1$". *Proceedings of the National Academy of Sciences, India, Sect. A.* 14, (1944), 101–105.

115. (With D. Singh) "A perfect difference set of order 18". *Mathematics Student.* 12, (1944), 85.

116. (With D. B. Lahiri, R. C. Bose and C. R. Rao) "On the integral order, (mod *p*) of quadratics $x^2 + ax + b$ with applications to the construction of minimum functions for $GF(p^2)$ and to some number theory results". *Bulletin of the Calcutta Mathematical Society.* 36, (1944), 153–174.

117. "On the difference sets". *Journal of the Indian Mathematical Society* (N. S.). 9, (1945), 28–31.

118. (With R. C. Bose and C. R. Rao) "Minimum functions in Galois fields". *Proceedings of the National Academy of Sciences, India, Sect. A.* 15, (1945), 191–192.

119. (With R. C. Bose and C. R. Rao) "On the roots of a well-known congruence". *Proceedings of the National Academy of Sciences, India, Sect. A.* 15, (1945), 193.

120. (With T. Vijayaraghavan) "Short proofs of theorems of Bose and Singer". *Proceedings of the National Academy of Sciences, India, Sect. A.* 15, (1945), 194.

121. (With D. Singh) "A perfect difference set of order 18". *Proceedings of the Lahore Philosophical Society.* 7, (1945), no. 1, 52.

122. (With R. C. Bose and C. R. Rao) "A chain of congruences". *Proceedings of the Lahore Philosophical Society.* 7, (1945), no. 1, 53.

123. (With R. C. Bose) "On a method of constructing a cyclic subgroup of order $p + 1$ of the group of linear fractional transformation mod *p*". *Proceedings of the Lahore Philosophical Society.* 7, (1945), no. 1, 53.

124. (With R. C. Bose) "On a method of constructing a cyclic subgroup of order $p + 1$ of the group of linear fractional transformation mod *p*". *Science and Culture.* 10, (1945), 558.

125. "On quintic equations soluble by radicals". *Mathematics Student.* 13, (1945), 84.

126. (With R. C. Bose) "On the construction of affine difference sets". *Bulletin of the Calcutta Mathematical Society.* 37, (1945), 107–112.

127. "Outline of a new method for proving results of elliptic function theory (such as identities of the Ramanujan–Rademacher–Zuckermanntype)". *Proceedings of the Lahore Philosophical Society.* 7, (1945), no. 2, 54–55.

128. "A formula similar to Jacobsthal's for the explicit value of *x* in $p = x^2 + y^2$ where *p* is a prime of the form $4k + 1$". *Proceedings of the Lahore Philosophical Society.* 7, (1945), no. 2, 56–57.

129. "The cubic character of 2 (mod p)". *Proceedings of the Lahore Philosophical Society.* 7, (1945), no. 2, 58.

130. "A proof of Euler's result". *Proceedings of the Lahore Philosophical Society.* 7, (1945), no. 2, 59.

131. (With A. R. Nazir) "Numbers representable by a Ternary quadratic form II". *Mathematics Student.* 14, (1946), 23.

132. (With R. P. Bambah) "Some new congruence properties of Ramanujan's function $\tau(n)$". *Mathematics Student.* 14, (1946), 24–26.

133. (With R. P. Bambah) "On integer cube roots of the unit matrix". *Science and Culture.* 12, (1946), 105.

134. "A note on multiplicative functions". *Proceedings of the National Institute of Sciences, India.* 12, (1946), 429–439.

135. (With R. P. Bambah) "A congruence property of Ramanujan's function $\tau(n)$". *Mathematics Student.* 14, (1946), 431–432.

136. (With R. P. Bambah) "On a function of Ramanujan". *Proceedings of the National Institute of Sciences, India.* 12, (1946), 433.

137. (With R. P. Bambah) "A note on Ramanujan's function $\tau(n)$". *Quarterly Journal of Mathematics.* Oxford, Ser. 18, (1947), 122–123.

138. (With R. P. Bambah, H. Gupta and D. B. Lahiri) "Congruence properties of Ramanujan's function $\tau(n)$". *Quarterly Journal of Mathematics.* Oxford, Ser. 18, (1947), 143–146.

139. (With R. P. Bambah, H. Gupta) "A congruence property of Ramanujan's function $\tau(n)$". *Bulletin of the American Mathematical Society.* 53, (1947), 766–767.

140. (With R. P. Bambah) "A new congruence property of Ramanujan's function $\tau(n)$". *Bulletin of the American Mathematical Society.* 53, (1947), 768–769.

141. "Proof of a theorem of Lerch and P. Kesava Menon". *Mathematics Student.* 15, (1947), 4.

142. "A theorem in analytic number theory". *Proceedings of the National Institute of Sciences, India.* 13, (1947), 97–99.

143. "On a theorem of Walfisz". *Journal of the London Mathematical Society.* 22, (1947), 136–140.

144. (With R. P. Bambah) "The residue of Ramanujan's function $\tau(n)$ to the modulus 2^8". *Journal of the London Mathematical Society.* 22, (1947), 140–147.

145. (With R. P. Bambah) "On numbers which can be expressed as a sum of two squares". *Proceedings of the National Institute of Sciences, India,* 13, (1947), 101–103.

146. "Modular equations as solutions of algebraic differential equations of the sixth order". *Proceedings of the National Institute of Sciences, India.* 13, (1947), 169–170.

147. (With R. P. Bambah) "On the sign of the Gaussian sum". *Proceedings of the National Institute of Sciences, India.* 13, (1947), 175–176.

148. "On an unsuspected real zero of Epstein's seta function". *Proceedings of the National Institute of Sciences, India.* 13, (1947), 177.

149. "On the class-number of the corpus $P(\sqrt{-k})$". *Proceedings of the National Institute of Sciences, India,* 13, (1947), 197–200.

150. "On a problem of analytic number theory". *Proceedings of the National Institute of Sciences, India.* 13, (1947), 231–233.

151. (With R. P. Bambah) "On integer roots of the unit matrix". *Proceedings of the National Institute of Sciences, India.* 13, (1947), 241–246.

152. (With R. P. Bambah) "Congruence properties of Ramanujan's function $\tau(n)$". *Bulletin of the American Mathematical Society.* 53, (1947), 950–955.

153. (With T. Vijayaraghavan) "On the largest prime divisors of numbers". *Journal of the Indian Mathematical Society* (N. S.). 11, (1947), 31–37.

154. (With T. Vijayaraghavan) "On complete residue sets". *Quarterly Journal of Mathematics.* Oxford, Ser. 19, (1948), 193–199.

155. "An improvement of a theorem of Linnik and Walfisz". *Proceedings of the National Institute of Sciences, India.* 15, (1949), 81–84.

156. "Improvement of a theorem of Linnik and Walfisz". *Proceedings of the London Mathematical Society.* (2) 50, (1949), 423–429.

157. (With J. Todd) "The density of reducible integers". *Canadian Journal of Mathematics.* 1, (1949), 297–299.
158. "Prime numbers and allied topics". *Proceedings of the 36th Indian Science Congress (Allahabad, 1949), Part II: Presidential Addresses.* pp. 1–7, Indian Science Congress Association, Calcutta, 1949.
159. "On difference sets". *Proceedings of the National Academy of Sciences.* U.S.A., 35, (1949), 92–94.
160. "The last entry in Gauss's diary". *Proceedings of the National Academy of Sciences.* U.S.A., 35, (1949), 244–246.
161. (With A. Selberg) "On Epstein's seta-function (I)". *Proceedings of the National Academy of Sciences.* U.S.A., 35, (1949), 371–374.
162. (With N. C. Ankney) "The class number of the cyclotomic field". *Proceedings of the National Academy of Sciences.* U.S.A., 35, (1949), 529–532.

Hansraj Gupta (HG)

1. "Sum of products of the first n natural numbers taken r at a time". *Journal of the Indian Mathematical Society.* 19, (1931), 1–6.
2. "On numbers in medial progression". *Journal of the Indian Mathematical Society.* 19, (1932), 203–214.
3. "A problem in diophantine analysis". *American Journal of Mathematics.* 56, (1934), 269–274.
4. "A table of partitions". *Proceedings of the London Mathematical Society.* 39, (1935), 142–149.
5. "On G-functions in general". *Mathematics Student.* 3, (1935), 50–55.
6. "Congruence properties of G-functions". *Proceedings of the Edinburgh Mathematical Society.* Series 2, 4, (1935), 61–66.
7. "On a theorem of Gauss". *Proceedings of the Edinburgh Mathematical Society.* Series 2, 4, (1935), 118–120.
8. "A generalization of a theorem of Wolstenholme". *Mathematical Notes.* Edinburgh Mathematical Society, 29, (1935), xi–xiii.
9. "On the p-potency of $G(n, r)$". *Proceedings of the Indian Academy of Science, Sect. A.* 1, (1935), 620–622.
10. "On the p-potency of $G(p^\mu - 1, r)$". *Proceedings of the Indian Academy of Science, Sect. A.* 2, (1935), 199–202.
11. "Decompositions into squares of primes". *Proceedings of the Indian Academy of Science, Sect. A.* 1, (1935), 789–794.
12. "Decompositions into cubes of primes". *Journal of the London Mathematical Society.* 10, (1935), 275.
13. "On a Brocard–Ramanujan problem". *Mathematics Student.* 3, (1935), 71.
14. "Two more perfect numbers". *American Mathematical Monthly.* 42, (1935), 163–174.
15. "On linear quotient sequences". *Mathematics Student.* 3, (1935), 132–137.
16. "On partitions of n". *Journal of the London Mathematical Society.* 11, (1936), 278–280.
17. "On the numbers of Ward and Bernoulli". *Proceedings of the Indian Academy of Science, Sect. A.* 3, (1936), 193–200.
18. "On a conjecture of Ramanujan". *Proceedings of the Indian Academy of Science, Sect. A.* 4, (1936), 625–629.
19. "Minimum partitions into specified parts". *American Journal of Mathematics.* 58, (1936), 573–576.
20. "Decompositions into cubes of primes (II)". *Proceedings of the Indian Academy of Science, Sect. A.* 4, (1936), 216–221.
21. "On sums of powers". *Proceedings of the Indian Academy of Science, Sect. A.* 4, (1936), 571–574.
22. "On the Diophantine equation $m^2 = n! + 1$". *American Mathematical Monthly.* 43, (1936), 32–34.
23. "A table of partitions (II)". *Proceedings of the London Mathematical Society.* 42, (1937), 546–549.

24. "A note on Wilson's quotient". *Mathematics Student*. 5, (1937), 176–177.
25. "On a conjecture of Chowla". *Proceedings of the Indian Academy of Science, Sect. A*. 5, (1937), 381–384.
26. "Decompositions of primes into cubes". *Tohoku Mathematical Journal*. 43, (1937), 11–16.
27. "On squares in arithmetical progressions". *Mathematics Student*. 5, (1937), 111.
28. "A generalisation of Leudesdorf's theorem". *Proceedings of the Indian Academy of Science, Sect. A*. 7, (1938), 390–392.
29. "On a property of self-sufficient sets". *Mathematics Student*. 6, (1938), 73–74.
30. "Waring's theorem for powers of primes". *Journal of the Indian Mathematical Society*. 3, (1938), 136–145.
31. "Congruence properties of self-contained balanced sets". *Proceedings of the Edinburgh Mathematical Society*. 6, (1938), 1–3.
32. "Note on Dirichlet's L-functions". *Journal of the Indian Mathematical Society*. 3, (1938), 198–199.
33. "Another generalisation of Leudesdorf's theorem". *Journal of the London Mathematical Society*. 14, (1939), 86–88.
34. "Analogues of Bauer's theorems". *Proceedings of the Indian Academy of Science, Sect. A*. 9, (1939), 396–398.
35. "On a problem of arrangements". *Proceedings of the Indian Academy of Science, Sect. A*. 9, (1939), 399–403.
36. "A note on some Diophantine equations". *Mathematics Student*. 7, (1939), 29–30.
37. "On a table of values of $L(n)$". *Proceedings of the Indian Academy of Science, Sect. A*. 12, (1940), 407–409.
38. "Waring's problem for powers of primes II". *Journal of the Indian Mathematical Society*. 4, (1940), 71–79.
39. "On the absolute weight of an integer". *Proceedings of the Indian Academy of Science, Sect. A*. 12, (1940), 60–62.
40. "A problem in combinations". *Mathematics Student*. 8, (1940), 131–132.
41. "On the extraction of the square-root of surds". *Proceedings of the Benares Mathematical Society* (N. S.). 2, (1940), 33–37.
42. "Some properties of generalised combinatory functions". *Journal of the Indian Mathematical Society* (N. S.). 5, (1941), 27–31.
43. "Some idiosyncratic numbers of Ramanujan". *Proceedings of the Indian Academy of Science, Sect. A*. 13, (1941), 519–520.
44. "On numbers of the form $4^a(8b + 7)$". *Journal of the Indian Mathematical Society* (N. S.). 5, (1941), 192–202.
45. "An important congruence". *Proceedings of the Indian Academy of Science, Sect. A*. 13, (1941), 85–86.
46. (F. C. Auluck and S. Chowla) "On the maximum value of the number of partitions of n into k parts". *Journal of the Indian Mathematical Society* (N. S.). 6. (1942), 105–112.
47. "A formula in partitions". *Journal of the Indian Mathematical Society* (N. S.). 6. (1942), 115–117.
48. "On an asymptotic formula in partitions". *Proceedings of the Indian Academy of Science, Sect. A*. 16, (1942), 101–102.
49. "An inequality in partitions". *Journal of Bombay University*. 11, (1942), 16–18.
50. "On the class-numbers in binary quadratic forms". *Univ. Nac. Tucuman Rivista A*. 3, (1942), 283–299.
51. "On the maximum values of $p_k(n)$ and $\pi_k(n)$". *Journal of the Indian Mathematical Society* (N. S.). 7, (1943), 72–75.
52. "On residue chains". *Mathematics Student*. 11, (1943), 54–55.
53. "Congruence properties of $\tau(n)$". *Proceedings of the Benares Mathematical Society* (N. S.). 5, (1943), 17–22.
54. "A formula for $L(n)$". *Journal of the Indian Mathematical Society* (N. S.). 7, (1943), 68–71.

55. "A congruence relation between $\tau(n)$ and $\sigma(n)$". *Journal of the Indian Mathematical Society* (N. S.). 9, (1945), 59–60.
56. "Congruence properties of $\sigma(n)$". *Mathematics Student*. 13, (1945), 25–29.
57. "A solution of the general quartic"*Mathematics Student*. 13, (1945), 31.
58. "A note on the parity of $p(n)$". *Journal of the Indian Mathematical Society* (N. S.). 10, (1946), 32–33.
59. "An asymptotic formula in partitions". *Journal of the Indian Mathematical Society* (N. S.). 10, (1946), 73–76.
60. "A congruence property of $\tau(n)$". *Proceedings of the Indian Academy of Science, Sect. A*. 24, (1946), 441–442.
61. "On n-th power residues". *Quarterly Journal of Mathematics*. Oxford, 18, (1947), 253–256.
62. (With R. P. Bambah, S. Chowla and D. B. Lahiri) "Congruence properties of Ramanujan's function $\tau(n)$". *Quarterly Journal of Mathematics*. Oxford, 18, (1947), 143–146.
63. (With R. P. Bambah and S. Chowla) "A congruence property of Ramanujan's function $\tau(n)$". *Bulletin of the American Mathematical Society*. 53, (1947), 766–767.
64. "A table of values of $N_3(t)$". *Proceedings of the National Institute of Sciences, India*. 13, (1947), 35–63.
65. "A table of values of $\tau(n)$". *Proceedings of the National Institute of Sciences, India*. 13, (1947), 201–206.
66. "On Kemmer's identity in combinatory functions". *Mathematics Student*. 15, (1947), 93–95.
67. "The vanishing of Ramanujan's function $\tau(n)$". *Current Science*. 17, (1948), 180.
68. "On $N_q(r)$ in the Tarry–Escott problem". *Proceedings of the National Institute of Sciences, India*. 14, (1948), 335–336.
69. "On a conjecture of Miller". *Journal of the Indian Mathematical Society* (N. S.). 13, (1949), 85–90.
70. "A solution of the Tarry–Escott problem of degree r". *Proceedings of the National Institute of Sciences, India*. 15, (1949), 37–39.
71. "On $\tau(n)$ modulo 49". *Current Science*. 18, (1949), 119.
72. "Some conjectures in number theory". *Current Science*. 18, (1949), 241.
73. "Tables of distributions". *Research Bulletin Panjab University*. 2, (1950), 13–44.
74. "A table of values of Liouville's function $L(t)$". *Research Bulletin of Panjab University*. 3, (1950), 45–63.
75. "On a problem of Erdös". *American Mathematical Monthly*. 57, (1950), 326–329.
76. "A generalization of the partition function". *Proceedings of the National Institute of Sciences, India*, 17, (1951), 231–238.
77. "Analogues of some $\mu(n)$ theorems". *Mathematics Student*. 19, (1951), 19–24.
78. "A note on sums of powers". *Mathematics Student*. 19, (1951), 117.
79. "A table of values $N_2(t)$". *Research Bulletin of East Panjab University*. 20, (1952), 13–93.
80. "A generalization of the Möbius function". *Scripta Mathematica*. 19, (1953), 121–126.
81. "Non-cyclic sets of points". *Proceedings of the National Institute of Sciences, India*. 19, (1953), 315–316.
82. "On a generating function in partition theory". *Proceedings of the National Institute of Sciences, India, Part A*. 20, (1954), 582–586.
83. (With M. S. Cheema and O. P. Gupta) "On Möbius means". *Research Bulletin of Panjab University*. 42, (1954), 1–17.
84. "A summation problem". *Mathematics Student*. 22, (1954), 105–107.
85. "On triangular numbers in arithmetical progression". *Mathematics Student*. 22, (1954), 141–143.
86. "Partitions into distinct primes". *Proceedings of the National Institute of Sciences, India, Part A*. 21, (1955), 185–187.
87. "Partitions in general". *Research Bulletin of Panjab University*. 67, (1955), 31–38.
88. "Some properties of quadratic residues". *Mathematics Student*. 23, (1955), 105–107.
89. "Partitions in terms of combinatory functions". *Research Bulletin of Panjab University*. 94, (1956), 153–159.

90. "Certain averages connected with partitions". *Research Bulletin of Panjab University.* 124, (1957), 427–430.
91. "Partition of *j*-partite numbers into *k*summands". *Journal of the London Mathematical Society.* 33, (1958), 403–405.
92. "Partition of *j*-partite numbers". *Research Bulletin of Panjab University.* 146, (1958), 119–121.
93. "Graphic representation of a partition of a *j*-partite number". *Research Bulletin of Panjab University* (N. S.). 10, (1959), 189–196.
94. "On the partition of *j*-partite numbers". *Proceedings of the National Institute of Sciences, India, Part A.* 27, (1961), 579–587.
95. "An inequality for $P(N_j, k)$". *Research Bulletin of Panjab University* (N. S.). 13, (1962), 173–178.
96. "Partition of *j*-partite numbers". *Mathematics Student.* 31, (1963), 179–186.
97. (With A. M. Vaidya) "The number of representations of a number as a sum of two squares". *American Mathematical Monthly.* 70, (1963), 1081–1082.
98. "On the coefficients of the powers of Dedekind's modular form". *Journal of the London Mathematical Society.* 39, (1964), 433–440.
99. "A congruence property of Euler's φ-function". *Journal of the London Mathematical Society.* 39, (1964), 303–306.
100. "On a problem in matrices". *Proceedings of the National Institute of Sciences, India, Part A.* 30, (1964), 556–560.
101. "Classes of divisors modulo 24". *Proceedings of the National Institute of Sciences, India, Part A.* 30, (1964), 758–766.
102. "An identity". *Research Bulletin of Panjab University* (N. S.). 15, (1964), 347–349.
103. "Partitions: a survey". *Mathematics Student.* 32, (1964), 1–19.
104. "On some problems suggested by mathematical tables". *Presidential Address at the Section of Mathematics.* Indian Science Congress, Calcutta, 1964–65, 9–22.
105. (With Harsh Anand and Vishwa Chander Dumir) "A combinatorial distribution problem". *Duke Mathematical Journal.* 33, (1966), 757–769.
106. "Pseudo-primitive-roots and indices". *Research Bulletin of Panjab University* (N. S.). 18, (1966), 251–252.
107. "A sum involving the Möbius function". *Proceedings of the American Mathematical Society.* 19, (1968), 445–447.
108. (With Cheema, M. S.) "The maxima of $P_r(n_1, n_2)$". *Mathematical Computations.* 22, (1968), 199–200.
109. "Number of topologies on a finite set". *Research Bulletin of Panjab University* (N. S.). 19, (1968), 231–241.
110. "An arithmetical sum". *Indian Journal of Mathematics.* 10, (1968), 83–86.
111. "Enumeration of symmetric matrices". *Duke Mathematical Journal.* 35, (1968), 653–659.
112. "On a problem in parity". *Indian Journal of Mathematics.* 11, (1969), 157–163.
113. "Highly restricted partitions". *Journal Res. National Bureau of Standards.* 73B, (1969), 329–350.
114. "Three combinatorial problems". *Research Bulletin of Panjab University* (N. S.). 20, (1969), 443–448 (1970)
115. "Partitions:asurvey". *Journal Res. National Bureau of Standards.* 74B, (1970), 1–29.
116. "Products of parts in partitions into primes". *Research Bulletin of Panjab University* (N. S.). 21, (1970), 251–253.
117. "Meanings of non-associative expressions". *Research Bulletin of Panjab University* (N. S.). 21, (1970), 255–256.
118. "Chains of quadratic residues". *Math. Comp.* 25, (1971), 379–382.
119. "Proof of the Churchhouse conjecture concerning binary partitions". *Proceedings of the Cambridge Philosophical Society.* 70, (1971), 53–56.
120. "On Sylvester's theorem in partitions". *Indian Journal of Pure and Applied Mathematics.* 2, (1971), 740–748.

121. "On partitions of n into k summands". *Proceedings of the Edinburgh Mathematical Society.* 17(2), (1970/71), 337–339.

122. "On the enumeration of symmetric matrices". *Duke Mathematical Journal.* 38, (1971), 709–710.

123. "Partial fractions in partition theory". *Research Bulletin of Panjab University* (N. S.). 22, (1971), 23–25.

124. "A combinatorial identity". *Indian Journal of Mathematics.* 13, (1971), 139–140.

125. "A problem in permutations and Stirling's numbers". *Mathematics Student.* 39, (1971), 341–345, (1972).

126. (With Srinivasan and Seshadri) "Cycles of quadratic congruences". *Research Bulletin of Panjab University* (N. S.). 22, (1971), 401–404 (1972).

127. (With Srinivasan and Seshadri) "The number of 3 by 3 magic matrices". *Research Bulletin of Panjab University* (N. S.). 22, (1971), 525–526 (1972).

128. "On m-ary partitions". *Proceedings of the Cambridge Philosophical Society.* 71, (1972), 343–345.

129. "The Legendre and Jacobi symbols for k-ic residues". *Journal of Number Theory.* 4, (1972), 219–222.

130. "Two theorems in partitions". *Indian Journal of Mathematics.* 14, (1972), 7–8.

131. "Restricted solid partitions". *Journal of Combinatorial Theory, Ser A.* 13, (1972), 140–144.

132. "A simple proof of the Churchhouse conjecture concerning binary partitions". *Indian Journal of Pure and Applied Mathematics.* 3, (1972), 791–794.

133. "Partitions of j-partite numbers into twelve or a smaller number of parts". *Mathematics Student.* 40, (1972), 401–441 (1974).

134. "Product decomposition of complete residue sets". *Indian Journal of Mathematics.* 15, (1973), No. 3, 191–195.

135. "The combinatorial recurrence". *Indian Journal of Pure and Applied Mathematics.* 4, (1973), 529–532.

136. "A restricted Diophantine equation". *Journal of the Indian Mathematical Society* (N. S.). 37, (1973), 71–77, (1974).

137. "Frequency distribution of digits over the first N natural numbers". *Mathematics Student.* 41, (1973), 447–448 (1974).

138. (With Bhattacharjya, G.) "A sum involving the greatest integer function". *Research Bulletin of Panjab University* (N. S.). 24, (1973), 51–53, (1977).

139. "Ramanujan's ternary quadratic form $x^2 + y^2 + 10z^2$". *Research Bulletin of Panjab University* (N. S.). 24, (1973), 57 (1977).

140. (With Nath, G. Baikunth) "Enumeration of stochastic cubes". *Indian Journal of Pure and Applied Mathematics.* 4, (1973), 545–567.

141. "A partition theorem of Subba Rao". *Canadian Mathematical Bulletin.* 17, (1974), 121–123.

142. "Primary factorizations of complete residue sets". *Indian Journal of Pure and Applied Mathematics.* 5, (1974), 1085–1092.

143. "Some sequences with distinct sums". *Indian Journal of Pure and Applied Mathematics.* 5, (1974), 1093–1109.

144. "On the Diophantine equation $x^n = y_1 y_2 \cdots y_m$". *Univ. Beograd. Publ. Elektronetehn. Fak. Ser. Mat. Fiz.* No. 461–No. 497 (1974), 139–142.

145. "Magic partitions". *Indian Journal of Pure and Applied Mathematics.* 6, (1975), 1159–1166.

146. "Partitions of n into divisors of m". *Indian Journal of Pure and Applied Mathematics.* 6, (1975), 1276–1286.

147. "Three-dimensional models of partitions of bipartite numbers". *Indian Journal of Pure and Applied Mathematics.* 6, (1975), 1287–1308.

148. "A technique in partitions". *Univ. Beograd. Publ. Elektronetehn. Fak. Ser. Mat. Fiz.* No. 498–No. 541 (1975), 73–76.

149. "An algorithm for finding the prime divisors of (k^n)". *Univ. Beograd. Publ. Elektronetehn. Fak. Ser. Mat. Fiz.* No. 498–No. 541 (1975), 77–83.

150. "A direct method of obtaining Farey–Fibonacci sequences". *Fibonacci Quarterly*. 14, (1976), 389–391.

151. "A direct proof of the Churchhouse conjecture concerning binary partitions". *Indian Journal of Mathematics*. 18, (1976), 1–5.

152. "Combinatorial proof of a theorem on partitions into an even or odd number of parts". *Journal of Combinatorial Theory*. Ser A, 21, (1976), 100–103.

153. "Use of inverse function in summation". *Univ. Beograd. Publ. Elektronetehn. Fak. Ser. Mat. Fiz.* No. 544–No. 576 (1976), 101–102.

154. (With Erdös, P. and Khare, S. P.) "On the number of distinct prime divisors of (k^n)". *Utilitus Math*. 10, (1976), 51–60.

155. "Partitions embedded in a rectangle". *Utilitus Math*. 10, (1976), 229–240.

156. "Magic partitions (I)". *Mathematics Student*. 45, (1977), 58–62.

157. (With Khare, S. P.) "On (k^{k2}) and the product of first $k$primes". *Univ. Beograd. Publ. Elektronetehn. Fak. Ser. Mat. Fiz.* No. 577–No. 598 (1977), 25–29.

158. "The rank vector of a partition". *Fibonacci Quarterly*. 16, (1978), 548–552.

159. "The Andrews formula for Fibonacci numbers". *Fibonacci Quarterly*. 16, (1978), 552–555.

160. "Finite differences of the partition function". *Math. Comp*. 32, (1978), No. 144, 1241–1243.

161. "A new look at the permutations of the first n natural numbers". *Indian Journal of Pure and Applied Mathematics*. 9, (1978), 600–631.

162. "Report on the solutions of the Diophantine equation". *Mathematics Student*. 46, (1978), 342–350.

163. "On sums of factors of m". *Mathematics Student*. 46, (1978), 379–391.

164. "Powers of 2 and sums of distinct powers of 3". *Univ. Beograd. Publ. Elektronetehn. Fak. Ser. Mat. Fiz.* No. 602–No. 633 (1978), 151–158 (1979).

165. (With Pleasants, P. A. B.) "Partitions into powers of m". *Indian Journal of Pure and Applied Mathematics*. 10, (1979), 655–694.

166. "Enumeration of incongruent cyclic k-gons". *Indian Journal of Pure and Applied Mathematics*. 10, (1979), 964–999.

167. (With Singh, Kuldip) 'The largest r for which $(n + k)!/n!(k + r)!$ is an integer". *Indian Journal of Pure and Applied Mathematics*. 10, (1979), 1249–1265.

168. "A note on the Stein–Waterman sequences". *Indian Journal of Pure and Applied Mathematics*. 11, (1980), 147–148.

169. "On cycles of k integers". *Indian Journal of Pure and Applied Mathematics*. 11, (4), (1980), 527–545.

170. "Euler's totient function and its inverse'". *Indian Journal of Pure and Applied Mathematics*. 12, (1981), 22–30.

171. "On permutation-generating strings and rosaries". *Combinatorics and Graph Theory*. Lecture Notes in Mathematics, 885, (1981), 272–275, Springer.

172. "On a partition-problem of Erdös". *Indian Journal of Pure and Applied Mathematics*. 12, (1981), 1293–1298.

173. The binary system and distribution matrices". *Indian Journal of Pure and Applied Mathematics*. 12, (1981), 1408–1419.

174. "Multi-scalar systems and distribution matrices". *Indian Journal of Pure and Applied Mathematics*. 13, (1982), 127–147.

175. "Diophantine equations in partitions". *Math. Comp*. 42, (1984), 225–229.

176. "Generators of primary nasty numbers". *National Academy of Science, Letters*. 7, (1984), No. 9, 289–290.

177. "Two formulae for $L(n)$". *Indian Journal of Pure and Applied Mathematics*. 15, (9), (1984), 957–961.

178. (With Kuldip Singh) "On k-triad sequences". *International Journal of Mathematics and Mathematical Sciences*. 8, (1985), 799–804.

R. P. Bambah (RPB)

Publications made in the twentieth century only have been listed.

1. (With S. Chowla) "On integer roots of the unit matrix". *Science and Culture.* 12, (1946), 105.
2. "On the complete primitive residue sets". *Bulletin of the Calcutta Mathematical Society.* 38, (1946), 113–116.
3. "Two congruence properties of Ramanujan's function $\tau(n)$". *Journal of the London Mathematical Society.* 21, (1946), 91–93.
4. (With S. Chowla) "A congruence property of Ramanujan's function $\tau(n)$". *Proceedings of the National Institute of Sciences, India.* 12, (1946), 431–432.
5. (With S. Chowla) "On a function of Ramanujan". *Proceedings of the National Institute of Sciences, India.* 12, (1946), No. 8, 1.
6. (With S. Chowla) "On a function of Ramanujan". *Proceedings of the National Institute of Sciences, India.* 12, (1946), 433.
7. "On integer cube roots of the unit matrix". *Mathematics Student.* 14, (1946), 69–70 (1948)
8. (With S. Chowla) "Some new congruence properties of Ramanujan's function $\tau(n)$". *Mathematics Student.* 14, (1946), 24–26.
9. (With S. Chowla) "A note on Ramanujan's function $\tau(n)$". *Quarterly Journal of Mathematics.* Oxford, Ser 18, (1947), 122–123.
10. (With S. Chowla) "Ramanujan's function:acongruence property". *Bulletin of the American Mathematical Society.* 53, (1947), 764–765.
11. (With S. Chowla and H. Gupta) "A congruence property of Ramanujan's function $\tau(n)$". *Bulletin of the American Mathematical Society.* 53, (1947), 766–767.
12. (With S. Chowla) "A new congruence property of Ramanujan's function $\tau(n)$". *Bulletin of the American Mathematical Society.* 53, (1947), 768–769.
13. (With S. Chowla, H. Gupta and D. B. Lahiri) "Congruence properties of Ramanujan's function". *Quarterly Journal of Mathematics*, Oxford, Ser. 18, (1947), 143–146.
14. (With S. Chowla) "Congruence properties of Ramanujan's function $\tau(n)$". *Bulletin of the American Mathematical Society.* 53, (1947), 950–955.
15. (With S. Chowla) "On numbers which can be expressed as a sum of two squares". *Proceedings of the National Institute of Science.* 13, (1947), 101–103.
16. (With S. Chowla) "On integer roots of the unit matrix". *Proceedings of the National Institute of Science.* 13, (1947), 241–246.
17. (With S. Chowla) "The residue of Ramanujan's function $\tau(n)$ to the modulus 2^8". *Journal of the London Mathematical Society.* 22, (1947), 140–147.
18. (With S. Chowla) "On the sign of the Gaussian sum". *Proceedings of the National Institute of Science.* 13, (1947), 175–176.
19. "On the geometry of numbers of non-convex star-regions with hexagonal symmetry". *Philosophical Transactions of the Royal Society.* London, Ser. A, 243, (1951), 431–462.
20. "Non-homogeneous binary cubic forms". *Proceedings of the Cambridge Philosophical Society.* 47, (1951), 457–460.
21. "Non-homogeneous binary quadratic forms I: two theorems of varnavides". *Acta Mathematica.* 86, (1951), 1–29.
22. "Non-homogeneous binary quadratic forms II: the second minimum of $(x + x_0)^2 - 7(y + y_0)^2$". *Acta Mathematica.* 86, (1951), 31–56.
23. (With C. A. Rogers) "Covering the planes with convex sets". *Journal of the London Mathematical Society.* 27, (1952), 304–314.
24. (With K. F. Roth) "A note on lattice coverings". *Journal of the Indian Mathematical Society* (N. S.). 16, (1952), 7–12.
25. (With H. Davenport) "The covering of n-dimensional space by spheres". *Journal of the London Mathematical Society.* 27, (1952), 224–229.
26. "On lattice coverings". *Proceedings of the National Institute of Sciences, India.* 19, (1953), 447–459.

27. "On polar reciprocal convex domains". *Proceedings of the National Institute of Sciences, India.* 20, (1954), 119–120.

28. "On lattice covering by spheres". *Proceedings of the National Institute of Sciences, India.* 20, (1954), 25–52.

29. "Lattice coverings with four-dimensional spheres". *Proceedings of the Cambridge Philosophical Society.* 50, (1954), 203–208.

30. "On polar reciprocal convex domains, addendum". *Proceedings of the National Institute of Sciences, India.* 20, (1954), 324–325.

31. "Four squares and a k-th power". *Quarterly Journal of Mathematics.* Oxford, Ser. (2) 5, (1954), 191–202.

32. (With K. Rogers) "An inhomogeneous minimum for non-convex star-regions with hexagonal symmetry". *Canadian Journal of Mathematics.* 7, (1955), 337–346.

33. "Polar reciprocal convex bodies". *Proceedings of the Cambridge Philosophical Society.* 51, (1955), 377–378.

34. "Divided cells". *Research Bulletin of Panjab University.* (1955), No. 81, 173–174.

35. "Maximal covering domains". *Proceedings of the National Institute of Sciences, India, Part A.* 23, (1957), 540–543.

36. "An analogue of a problem of Mahler". *Research Bulletin of Panjab University.* 109, (1957), 299–302.

37. "Some transference theorems in the geometry of numbers". *Montash. Math.* 62, (1958), 243–249.

38. "Some problems in the geometry of numbers". *Journal of the Indian Mathematical Society* (N. S.). 24, (1960), 157–172 (1961).

39. (With D. D. Joshi and I. S. Luthar) "Some lower bounds on the number of code points in a minimum distance binary code, I, II". *Information and Control.* 4, (1961), 313–319, 320–323.

40. (With I. S. Luthar and M. L. Madan) "On the existence of a certain type of basis in a totally real field". *Research Bulletin of Panjab University.* 12, (1961), 135–137.

41. "A note on the equation $ax^2 - by^2 - cz^2 = 0$". *Indian Journal of Mathematics.* 4, (1962), 11–12.

42. (With Rogers, C. A.) "On coverings with convex domains". *Acta Arithmetica.* 9, (1964), 191–207.

43. (With Woods, Allan and Zassenhaus, Hans) "Three proofs of Minkowski's second inequality in the geometry of numbers". *Journal of the Australian Mathematical Society.* 5, (1965), 453–462.

44. (With Woods, A. C) "On the minimal density of maximal packings of the plane by convex bodies"*Acta Mathematica Academy Science Hungary.* 19, (1968), 103–116.

45. (With Woods, A. C.) "Convex bodies with a covering property". *Journal of the London Mathematical Society.* 43, (1968), 53–56.

46. (With Woods, A. C.) "The covering constant for a cylinder". *Montash. Math.* 72, (1968), 107–117.

47. (With Woods, A. C.) "On minimal density of plane covering by circles". *Acta Mathematica Academy Science Hungary.* 19, (1968), 337–343.

48. "Packing and covering". *Mathematics Student.* 38, (1970), 133–138.

49. (With Woods, A. C.) "On plane coverings with convex domains". *Mathematika.* 18, (1971), 321–336.

50. (With Woods, A. C.) "The thinnest double covering of three-spheres". *Acta Arithmetica.* 18, (1971), 321–336.

51. (With Woods, A. C.) "On a problem of Danzer". *Pacific Journal of Mathematics.* 37, (1971), 295–301.

52. "Geometry of numbers, packing and covering and discrete geometry". Volume dedicated to the memory of V. Ramaswami Aiyar. *Mathematics Student.* 39, (1971), 117–129 (1972).

53. (With Woods, A. C.) "On a theorem of Dyson". Collection of articles dedicated to K. Mahler on the occasion of his seventieth birthday. *Journal of Number Theory.* 6, (1974), 422–433.

54. (With Woods, A. C.) "On the product of tree inhomogeneous linear forms". *Number Theory and Algebra.* pp. 7–18, Academic Press, New York, (1977).

55. (With Dumir, V. C. and Hans-Gill, R. J.) "Covering by star domains". *Indian Journal of Pure and Applied Mathematics.* 8, (1977), No. 3, 344–350.

56. (With Woods, A. C.) "Minkowski's conjecture for $n = 5$:atheorem of Skubenko". *Journal of Number Theory.* 12, (1980), No. 1, 27–48.

57. "Srinivasa Ramanujan medal lecture 1979:Number theory:many challenges, some achievements". *Proceedings of the Indian National Science Academy, Part A.* 46, (1980), No. 2, 109–118.

58. (With Dumir, V. C. and Hans-Gill, R. J.) "Positive values of nonhomogeneous indefinite quadratic forms". *Topics in Classical Number Theory.* Vol. I, II (Budapest, 1981), 111–170.

59. (With Sloane, N. J. A.) "On a problem of Ryškov concerning lattice coverings". *Acta Arithmetica.* 42, (1982/83), No. 1, 107–109.

60. (With Dumir, V. C. and Hans-Gill, R. J.) "On a conjecture of Jackson on nonhomogeneous qadratic forms". *Journal of Number Theory.* 16, (1983), No. 3, 403–419.

61. (With Dumir, V. C. and Hans-Gill, R. J.) "Positive values of Nonhomogeneous indefinite quadratic forms II". *Journal of Number Theory,* 18, (1984), No. 3, 313–341.

62. (With Dumir, V. C. and Hans-Gill, R. J.) "On an analogue of a problem of Mordell". *Studia Sci. Math. Hungar.* 21, (1986), No. 1–2, 135–142.

63. (With Woods, A. C.) "On a problem of G. Fejes Tóth". K. G. Ramanathan Memorial Issue. *Proceedings of the Indian Academy of Sciences, Mathematical Science.* 104, (1994), No. 1, 137–156.

64. "The conjectures of Minkowski, Watson and others'. *Mathematics Student.* 65, (1996), No. 1–4, 176–178.

65. (With Dumir, V. C. and Hans-Gill, R. J.) "Diophantine inequalities". *Proceedings of the National Academy of Sciences, India, Sect. A.* 68, (1998), No. 2, 101–114.

66. "Chowla, the mathematics man". *Mathematics Student.* 67, (1998), No. 1–4, 153–161.

67. "Packings and coverings". *Indian Journal of Pure and Applied Mathematics.* 30, (1999), No. 10, 1063–1072.

A. R. Rajwade (ARR)

Publications made in the twentieth century only have been listed.

1. "Arithmetic on curves with complex multiplication by $\sqrt{-2}$". *Proceedings of the Cambridge Philosophical Society.* 64, (1968), 659–672.

2. "Arithmetic on curves with complex multiplication by Eisentein integers". *Proceedings of the Cambridge Philosophical Society.* 65, (1969), 59–73.

3. "On rational primes p congruent to 1 (mod 3 or 5)". *Proceedings of the Cambridge Philosophical Society.* 66, (1969), 61–70.

4. "A note on the number of solutions N_p of the congruence $y^2 \equiv x^3 - Dx$ (modulo p)". *Proceedings of the Cambridge Philosophical Society.* 7, (1970), 603–605.

5. "On the congruence $y^2 \equiv x^5 - a$ (mod p)". *Proceedings of the Cambridge Philosophical Society.* 74, (1973), 473–475.

6. "The number of solutions of the congruence $y^2 \equiv x^6 - a$(mod p)". *Indian Journal of Pure and Applied Mathematics.* 4, (1973), 325–332.

7. (With Singh, Surjit) "The number of solutions of the congruence $y^2 \equiv x^4 - a$ (mod p)". *L'Enseignement Mathematique.* 20, (1974), 265–273.

8. "Notes on the congruence $y^2 \equiv x^5 - a$ (mod p)". *L'Enseignement Mathematique.* 21, (1975), 49–56.

9. "A note on stufe of qadratic fields". *Indian Journal of Pure and Applied Mathematics.* 6, (1975), 725–726.

10. "Note sur le theoreme destrios carres". *L'Enseignement Mathematique.* 22, (1976), 171–173.

11. "Some congruences in algebraic integers and rational integers". *Indian Journal of Pure and Applied Mathematics.* 7, (1976), 431–435.

12. (With Singh, Budh) "Determination of a unique solution of the quadratic partition for primes $p = 1 \pmod 7$". *Pacific Journal of Mathematics*. 72, (1977), 513–521.

13. "Some formulae for elliptic curves with complex multiplications". *Indian Journal of Pure and Applied Mathematics*. 8, (1977), 379–387.

14. "The Diophantine equation $y^2 = x(x^2 + 21\,Dx + 112\,D^2)$ and the conjectures of Birch and Swinnerton–Dyer". *Journal of Australian Mathematical Society*. 24, (1977), Ser A, 286–295.

15. (With Parnami, J. C. and Agrawal, M. K.) "The number of points on the curve $y^2 \equiv x^5 + a$ in F_q and applications to local ζ- function". *The Mathematics Student*. 48, (1980), 205–211.

16. (With Parnami, J. C. and Agrawal, M. K.) "A new proof of the leonard and Williams criterion for 3 to be a 7thpower". *Journal of the Indian Mathematical Society*, 45, (1981), 129–134.

17. (With Parnami, J. C. and Agrawal, M. K.) "A congruence relation between the coefficients of the Jacobi sum". *Indian Journal of Pure and Applied Mathematics*. 12, (1981), 804–806.

18. (With Parnami, J. C. and Agrawal, M. K.) "On the 4-power Stufe of a field". *Rendiconti del Circolo Mathematico di Palermo*, Serie II, 30, (1981), 245–254.

19. (With Parnami, J. C. and Agrawal, M. K.) "On some tigonometric Diophantine equations of the type $\sqrt{n} = c_1 \cos \pi d_1 + \cdots + c_2 \cos \pi d_2 + \cdots + c_1 \cos \pi d_1$". *Acta Mathematica Academiae Scientiarum Hungaricae*. 37 (4), (1981), 423 $\cdots$ 432.

20. (With Parnami, J. C. and Agrawal, M. K.) "On expressing $\sqrt{p}$ as a rational linear combination of cosines of angles which are rational multiples of π". *Annales Universitatis Scientiarum Budapestinensis de Ronaldo Eosvos nominatae, Sectio Mathematica*. 25, (1982), 31–40.

21. (With Parnami, J. C. and Agrawal, M. K.) "On expressing a quadratic irrational as a rational linear combination of roots of unity". *Annales Universitatis Scientiarum Budapestinensis de Ronaldo Eosvos nominatae, Sectio Mathematica*. 25, (1982), 41–51.

22. (With Parnami, J. C. and Agrawal, M. K.) "Triangles and cyclic quadrilaterals, with angles that are rational multiples of π and sides at most quadratic over the rationals". *The Mathematics Student*. 50, (1982), 79–93.

23. (With Parnami, J. C. and Agrawal, M. K.) "Jacobi sums and cyclotomic nmbers for a finite field". *Acta Arithmetica*. 41, (1982), 1–13.

24. (With Parnami, J. C.) "A new cubic character sum". *Acta Arithmetica*. 40, (1982), 347–356.

25. (With Rishi, D. B. and Parnami, J. C.) "Complex multiplication by $\frac{1}{2}\,(1 + \sqrt{-19})$". *Indian Journal of Pure and Applied Mathematics*. 14, (1983), 630–634.

26. (With Katre, S. A.) "On the Jacobsthal sum $\varphi_9(a)$ and the related sum $\psi_9\,(a)$". *Maths. Scand.*, 53, (1983), 193–202.

27. (With Pal, S.) "Power stufe of Galois fields". *Bulletin de la Societe Mathematique de Belgique*. 35, (1983), 123–130.

28. (With Parnami, J. C. and Agrawal, M. K.) "Criterion for 2 to be an l-thpower". *Acta Arithmetica*. 43, (1984), 361–365.

29. "Evaluation of a cubic character sum using the $\sqrt{-19}$ division points of the curve $y^2 = x^3 - 2^3$ $\cdot 19x + 2 \cdot 19^2$". *Journal Number Theory*. 2, 19, (1984), 184–194.

30. "On a conjecture of Williams". *Bulletin de la Societe Mathematique de Belgique*. 36, (1984), Ser. B, 1–4.

31. (With Rishi, D. B.) "On the integer property of the L-function for the elliptic curve with complex multiplication by the Gaussian integers". *Journal of the Indian Mathematical Society*. 48, (1984), 9–18.

32. (With Katre, S. A.) "Complete solution of the cyclotomic problem in F_q for any prime modulus $l, q = p^\alpha, p \equiv 1 \pmod l$". *Acta Arithmetica*. 45, (1985), 183–199.

33. (With Katre, S. A.) "Euler's criterion for quintic non-residues". *Canadian Journal of Mathematics*. 37, (1985), 1008–1024.

34. (With Katre, S. A.) "Unique determination of cyclotomic numbers of order five". *Manuscripta Math*. 53, (1985), 65–75.

35. (With Parnami, J. C. and Agrawal, M. K.) "On the fourth power stufe of p-adic completions of algebraic number fields". *Rend. Sem. Mat. Univers. Politcen. Torino*. 44, (1986), 141–153.

36. (With Katre, S. A.) "On the Jacobsthal sum $\varphi_4(a)$ and the related sum $\psi_8(a)$". *Annales Universitatis Scientiarum Budapestinensis de Rolando Eotovos Nominatae, Sectio Math*. 29, (1986), 3–7.

37. (With Katre, S. A.) "Jacobsthal sums of prime order". *Indian Journal of Pure and Applied Mathematics.* 17, (1986), 1345–1362.
38. (With Katre, S. A.) "Resolution of the sign ambiguity in the determination of the cyclotomic numbers of order four and the corresponding Jacobsthal sum". *Maths. Scanda.* 60, (1987), 52–62.
39. (With Parnami, J. C. and Agrawal, M. K.) "Some linear relations between certain binomial coefficients modulo a prime". *Bulletin de la Societe Mathematique de Belgique.* 41, (1989), 339–345.
40. (With Parnami, J. C. and Agrawal, M. K.) "Some identities involving character sums and their applications". *Journal of the Indian Mathematical Society.* 54, (1989), 125–132.
41. (With Parnami, J. C. and Agrawal, M. K.) "On the representation of -1 as a sum of fourth and sixth powers". *Annales Univ. Sci. Budapest.* 33, (1990), 43–47.
42. (With Parnami, J. C. and Agrawal, M. K.) "On the stufe of quartic fields". *Journal of Number Theory.* 38, (1991), 106–109.
43. "Pfister's work on sums of squares". *Number Theory.* Indian National Science Academy: Hindustan Book Agency. (1999), 325–349.

V. C. Dumir (VCD)

1. (With Hans-Gill, R. J.) "On positive values of non-homogeneous quarternary quadratic forms of type (1,3)". *Indian Journal of Pure and Applied Mathematics.* 12, (1981), 814–825.
2. (With Grover, V. K.) "Some asymmetric inequalities for non-homogeneous indefinite binary quadratic forms". *Journal of the Indian Mathematical Society* (N. S.). 50, (1986), 21–28.
3. (With Sehmi, R.) "Positive values of non-homogeneous indefinite quadratic forms of type (3,2)". *Indian Journal of Pure and Applied Mathematics.* 23, (1992), 812–853.
4. (With Sehmi, R.) "Positive values of non-homogeneous indefinite quadratic forms of signature 1". *Indian Journal of Pure and Applied Mathematics.* 23, (1992), 855–864.
5. (With Sehmi, R.) "Positive values of non-homogeneous indefinite quadratic forms of type (1, 4)". *Proceedings of the Indian Academy of Sciences, Mathematical Sciences.* 104, (1994), 557–579.
6. (With Hans-Gill, R. J. and Woods, A. C.) "Values of non-homogeneous indefinite quadratic forms". *Journal of Number Theory.* 47, (1994), 190–197.
7. (With Hans-Gill, R. J. and Sehmi, R.) "Positive values of non-homogeneous indefinite quadratic forms of type (2,4)". *Journal of Number Theory.* 55, (1995), 261–284.
8. (With Sehmi, R.) "Isolated minima of non-homogeneous indefinite quadratic forms of type (3,2) or (2,3)". *Journal of the Indian Mathematical Society* (N. S.). 61, (1995), 197–212.
9. (With Hans-Gill, R. J.) "The second minimum positive values of non-homogeneous ternary quadratic forms of type (1,2)". *Ranchi University Mathematical Journal.* 28, (1997), 65–75.

S. K. Khanduja (SKK)

She was S. K. Gogia before her marriage. Publications made in the twentieth century only have been listed here.

1. (With Luthar, I. S.) "Quadratic unramified extensions of $Q(\sqrt{d})$". *Journal Reine Angew. Math.* Band 298, (1978), 108–111.
2. (With Luthar, I. S.) "The Brauer–Siegel theorem for algebraic function fields". *Journal Reine Angew. Math.* Band 299/300, (1978), 28–37.
3. (With Luthar, I. S.) "Real characters of the ideal class group and the narrow ideal class group of $Q(\sqrt{d})$". *Colloq. Math.* 41, (1979), No. 1, 153–159.
4. (With Luthar, I. S.) "Norms from certain extensions of $F_q(T)$". *Acta Arithmetica.* 38, (1981), No. 4, 325–340.
5. "Certain quadratic unramified extensions of $F(T, \sqrt{D(T)})$ and theory of genera for quadratic extensions of $F_q(T)$". *Journal of the Indian Mathematical Society* (N. S.). 53, (1988), No. 1–4, 51–65.,

6. (With Garg, U.) "On extensions of valuations to simple transcendental extensions". *Proceedings of the Edinburgh Mathematical Society*. 32, (1989), No. 1, 147–156.

7. (With Garg, U.) "Rank 2 valuations of $K(x)$". *Mathematika*. 37 (1990), No. 1, 97–105.

8. (With Garg, U.) "On rank extensions of valuations". *Colloq. Math*. 59, (1990), No. 1, 25–29.

9. "A note on a result of James Ax". *Journal Algebra*. 140 (1991), No. 2, 360–361.

10. "Value groups and simple transcendental extensions". *Mathematika*. 38 (1991), No. 2, 381–385.

11. "Prolongations of valuations to simple transcendental extensions with given residue field and value group". *Mathematika*. 38 (1991), No. 2, 386–390.

12. (With Garg, U.) "On residually generic prolongations of a valuation to a simple transcendental extension". *Journal of the Indian Mathematical Society* (N. S.). 57, (1991), No. 1–4, 101–108.

13. "On valuations of $K(x)$". *Proceedings of the Edinburgh Mathematical Society*. (2), 35, (1992), No. 3, 419–426.

14. (With Garg, U.) "Residue fields of valued function fields of conics". *Proceedings of the Edinburgh Mathematical Society*. 36, (1993), No. 3, 469–478.

15. (With Garg, U.) "Prolongations of a Krull valuation to a simple transcendental extension". *Journal of the Indian Mathematical Society* (N. S.). 59, (1993), No. 1–4, 13–21.

16. "A uniqueness oroblem in simple transcendental extensions of valued fields". *Proceedings of the Edinburgh Mathematical Society*. 37, (1994), No. 1, 13–23.

17. "On value groups and residue fields of some valued function fields". *Proceedings of the Edinburgh Mathematical Society*. 37, (1994), No. 3, 445–454.

18. "On a result of James Ax". *Journal Algebra*. 172, (1995), No. 1, 147–151.

19. (With Saha, J.) "A uniqueness problem in valued function fields of conics". *Bulletin of the London Mathematical Society*. 28, (1996), No. 5, 455–462.

20. "On residually transcendental valued function fields of conics". *Glasgow Mathematical Journal*. 38, (1996), No. 2, 137–145.

21. "A note on residually transcendental prolongations with uniqueness property". *Journal Mathematics*. Kyoto University, 36, (1996), No. 3, 553–556.

22. "On extensions of valuations with prescribed value groups and residue fields". *Journal of the Indian Mathematical Society* (N. S.). 62, (1996), No. 1–4, 57–60.

23. (With Saha, J.) "On a generalization of Einstein's irreducibility criterion". *Mathematika*. 44, (1997), No. 1, 37–41.

24. "An independent theorem in simple transcendental extensions of valued fields". *Journal of the Indian Mathematical Society* (N. S.). 64, (1997), No. 1–4, 243–248.

25. "Valued function fields with given genus and residue fields". *Journal of the Indian Mathematical Society* (N. S.). 64, (1997), No. 1–4, 115–121.

26. "Tame fields and tame extensions". *Journal Algebra*. 201, (1998), 647–655.

27. (With Saha, J.) "The prime maximal ideals in $R[X]$, R a one-dimensional prefer domain". *Indian Journal of Pure and Applied Mathematics*. 29, (1998), 12, 1275–1279.

28. (With Saha, J.) "On invariants of elements over a Henselian field". *Journal of the Indian Mathematical Society* (N. S.). 65, (1998), No. 1–4, 127–132.

29. "On Krasner's constant". *Journal Algebra*. 213, (1999), No. 1, 225–230.

30. (With Garg, U.) "On a query of Adrian Wadsworth". *Indian Journal of Pure and Applied Mathematics*. 30, (1999), No. 9, 945–949.

31. (With Saha, J.) "Generalized Hensel's lemma". *Proceedings of the Edinburgh Mathematical Society*. (2), 42, (1999), No. 3, 469–480.

32. (With Saha, J.) "A generalized fundamental principle". *Mathematika*. 46, (1999), No. 1, 83–92.

S. A. Katre (SAK)

Publications made in the twentieth century only have been listed.

1. (With A. R. Rajwade) "On the Jacobsthal sum $\varphi_9(a)$ and the related sum $\psi_9(a)$". *Math. Scand*. 53 (1983), 193–202.

2. (With A. R. Rajwade) "Euler's criterion for quintic nonresidues". *Canadian Journal of Mathematics.* 37, (1985), 1008–1024.
3. "Jacobsthal sums in terms of quadratic partitions of a prime". *Proceedings, Conference in Number Theory, Ootacamund, India, 1984. Lecture Notes in Mathemati.* Springer Verlag, 1122, (1985), 153–162.
4. (With A. R. Rajwade) "Unique determination of cyclotomic numbers of order five". *Manuscripta Math.* 53, (1985), 65–75.
5. (With A. R. Rajwade) "Complete solution of the cyclotomic problem in F_q for any prime modulus l, $q = p^\alpha$, $p \equiv 1 \pmod{l}$". *Acta Arithmetica.* 45, (1985), 183–199.
6. (With A. R. Rajwade) "On the Jacobsthal sum $\varphi_4(a)$ and the related sum $\psi_8(a)$". *Annales Universitatis Scientiarum Budapestinensis de Rolando Eotovos Nominatae, Sectio Math.* 29, (1986), 3–7.
7. (With A. R. Rajwade) "Jacobsthal sums of prime order". *Indian Journal of Pure and Applied Mathematics.* 17, (1986), 1345–1362.
8. (With A. R. Rajwade) "Resolution of the sign ambiguity in the determination of the cyclotomic numbers of order four and the corresponding Jacobsthal sum". *Maths. Scanda.* 60, (1987), 52–62.
9. "Cyclotomic numbers and a conjecture of Snapper". *Indian Journal of Pure and Applied Mathematics.* 20, (1989), 99–103.
10. (With Sangita A. Khule) "A discriminant criterion for matrices over orders in algebraic number fields to be sums of squares". *Proceedings, Symposium on Algebra and Number Theory.* Kochi, Kerala, (1990), 31–38.
11. "On numbers of solutions of equations over finite fields". *Proceedings, Instructional School, Algebraic Number Theory.* Mumbai University, (1995), pp. 7.
12. (With V. V. Acharya) "Cyclotomic numbers of order 2l, l an odd prime". *Acta Arithmetica.* 69, (1995), 51–74.
13. (With Anuradha Narasimhan) "Explicit evaluation of cyclotomic numbers of prime order". *A. V. Prasad Memorial Volume of Ranchi University Mathematical Journal.* 28, (1997), 77–84.
14. (With Anuradha Narasimhan) "Number of points on the pojective curves $aY^l = bX^l + cZ^l$ and $aY^{2l} = bX^{2l} + cZ^{2l}$ defined over finite fields, l an odd prime". *Journal of Number Theory.* 77, No. 2, (1999), 288–313.

Madhu Raka (MR)

Publications made in the twentieth century only have been listed.

1. (With R. J. Hans-Gill) "An asymmetric inequality for non-homogeneous ternary quadratic forms". *Monatshefte. Maths.* (1979), 281–295.
2. (With R. J. Hans-Gill) "Inhomogeneous minimum of indefinite quadratic forms in five variables of type (3, 2) or (2, 3): a conjecture of Watson". *Monatshefte. Maths.* 88, (1979), 305–320.
3. (With R. J. Hans-Gill) "Positive values of inhomogeneous 5-ary quadratic forms of type (3, 2)". *Journal Australian Mathematical Society (Series A).* 29, (1980), 439–453.
4. (With R. J. Hans-Gill) "An inequality for indefinite ternary quadratic forms of type (2,1)". *Indian Journal of Pure and Applied Mathematics.* 11(2), (1980), 994–1006.
5. (With R. J. Hans-Gill) "Some inequalities for non-homogeneous quadratic forms". *Indian Journal of Pure and Applied Mathematics.* 11(1), January, (1980), 60–74.
6. (With R. J. Hans-Gill) "Inhomogeneous minimum of indefinite quadratic forms in five variables of type (4, 1) or (1, 4): a conjecture of Watson". *Indian Journal of Pure and Applied Mathematics.* 11(1), January, (1980), 75–91.
7. (With R. J. Hans-Gill) "An inequality for indefinite ternary quadratic forms of type (2, 1)". *Indian Journal of Pure and Applied Mathematics.* 11(8), August, (1980).
8. "Inhomogeneous minimum of indefinite quadratic forms of signature + 1 or −1". *Mathematics, Proceedings of the Cambridge Philosophical Society,* 89, (1981), 225–235.

9.   (With R. J. Hans-Gill) "Positive values of inhomogeneous quinary quadratic forms of type (4, 1)". *Journal Australian Mathematical Society (Series A)*. 31, (1981), 175–188.
10. "Inhomogeneous minimum of indefinite forms in six variables: a conjecture of Watson". *Mathematics, Proceedings of the Cambridge Philosophical Society*. 94, (1983), 1–8.
11. "On a conjecture of Watson". *Mathematics, Proceedings of the Cambridge Philosophical Society*. 94, (1983), 9–22.
12. (With V. K. Grover) "On inhomogeneous minimum of indefinite binary quadratic forms". *Acta Mathematica*. 167, (1991), 287–298.
13. "Inhomogeneous minimum of a class of ternary quadratic forms". *Journal Australian Mathematical Society, (Series A)*. 55, (1993), 334–354.
14. (With Urmila Rani) "Positive values of non-homogeneous indefinite quadratic forms of type (3, 1)". *Osterreich Wiss. Math. Natur. KI Vienna*. 203, (1994–1995), 175–197.
15. (With Urmila Rani) "Positive values of non-homogeneous indefinite quadratic forms of signature +2 or –2". *Osterreich Wiss. Math. Natur. KI Vienna*, 203, (1994–1995), 198–213.
16. (With Urmila Rani) "Positive values of inhomogeneous indefinite quadratic forms of type (2, 1)". *Hokkaido Mathematical Journal*. Japan, Vol. 25, (1996), 215–230.
17. (With Urmila Rani) "Positive values of non-homogeneous indefinite quadratic forms of type (1, 4)". *Proceedings of the Indian Academy of Sciences (Mathematical Science)*. Vol. 107 (4), November, (1997), 329–361.
18. (With Urmila Rani) "Inhomogeneous minima of a class of quarternary quadratic forms of type (2, 2)". *Proceedings of the American Mathematical Society, CONM Math*. Vol. 210, (1998), 275–298.
19. (With Urmila Rani) "The second minimum of inhomogeneous quaternary quadratic forms of type (3, 1) or (1, 3)". *Ranchi University Mathematical Journal*. 28, (1997), 5–34 (Dr. A. V. Prasad Memorial Volume).
20. (With Urmila Rani) "Inhomogeneous minima of a class of quarternary quadratic forms of type (3, 1)". *Indian Journal of Pure and Applied Mathematics*. 29(9), September, 1998, 889–908.

Section 3

K. Chandrasekharan (KC)

1.   (With Narasimhan, R.) "Sur l'ordre moyen de quelques fonctions arithmétiques". (French) *C. R. Acad. Sci. Paris*. 251, (1960), 1333–1335.
2.   (With Narasimhan, R.) "Hecke's functional equation and the average order of arithmetical functions". *Acta Arithmetica*. 6, (1960/1961), 487–503.
3.   (With Narasimhan, R.) "On Hecke's functional equation". *Bulletin of the American Mathematical Society*. 67, (1961), 182–185.
4.   (With Narasimhan, R.) "Hecke's functional equation and arithmetical identities". *Annals of Mathematics* (2). 74, (1961), 1–23.
5.   (With Narasimhan, R.) "Functional equations with multiple gamma factors and the average order of arithmetical functions". *Annals of Mathematics* (2). 76, (1962), 93–136.
6.   (With Narasimhan, R.) "The average order of arithmetical functions and the approximate functional equation for a class of Zeta-functions". *Rend. Mat. e Appl.* (5), 21, (1962), 354–363.
7.   (With Narasimhan, R.) "The approximate functional equation for a class of Zeta-functions" *Mathematical Annals*. 152, (1963), 30–64.
8.   (With Narasimhan, R.) "On the mean value of the error term for a class of arithmetical functions". *Acta Mathematica*. 112, (1964), 41–67.
9.   (With Narasimhan, R.) "On lattice points in a random sphere". *Bulletin of the American Mathematical Society*. 73, (1967), 68–71.
10. (With Narasimhan, R.) "Zeta-functions of ideal classes in quadratic fields and their zeros on the critical line". *Comment. Math. Helv.* 43, (1968), 18–30.
11. (With Narasimhan, R.) "An approximate reciprocity formula for some exponential sums". *Comment. Math. Helv.* 43, (1968), 296–310.

12. (With Narasimhan, R.) "Sommes exponentielles associées á un corps de nombres algébriques". (French) *C. R. Acad. Sci. Paris, Sér.* A-B, 287, (1978), no. 4, A 181–A 182.

K. G. Ramanathan (KGR)

1. "On demlo numbers". *Mathematics Student.* 9, (1941), 112–114.
2. "Congruence properties of $\sigma(n)$, the sum of divisors of n". *Mathematics Student.* 11, (1943), 33–35.
3. "Multiplicative arithmetic functions". *Journal of the Indian Mathematical Society.* 7, (1943), 111–116.
4. "On Ramanujan's trigonometrical sum $C_m(n)$". *Journal of Madras University.* Sect. B, 15, (1943), 1–9.
5. "Congruence properties of Ramanujan's function $\tau(n)$". *Proceedings of the Indian Academy of Science.* Sect. A 19, (1944), 146–148.
6. "Some applications of Ramanujan's trigonometrical sum $C_m(n)$". *Proceedings of the Indian Academy of Science.* Sect. A 20, (1944), 62–69.
7. "Congruence properties of $\sigma_a(n)$". *Mathematics Student.* 13, (1945), 30.
8. "Congruence properties of Ramanujan's function $\tau(n)$, II". *Journal of the Indian Mathematical Society.* 9, (1945), 55–59.
9. "Congruence properties of $\sigma_a(N)$". *Proceedings of the Indian Academy of Science.* Sect. A 25, (1947), 314–321.
10. "On the product of the elements in a finite Abelian group". *Journal of the Indian Mathematical Society.* 11, (1947), 44–48.
11. "Identities and congruences of the Ramanujan type". *Canadian Journal of Mathematics.* 2, (1950), 168–178.
12. "The theory of units of quadratic and Hermitian forms". *American Journal of Mathematics.* 73, (1951), 233–235.
13. "Abelian qadratic forms". *Canadian Journal of Mathematics.* 4, (1952), 352–368.
14. "Units of quadratic forms". *Annals of Mathematics.* 56, (1952), 1–10.
15. "A note on symplectic complements". *Journal of the Indian Mathematical Society.* 18, (1954), 115–125.
16. "The Riemann sphere in matrix spaces". *Journal of the Indian Mathematical Society.* 19, (1955), 121–125.
17. "Quadratic forms over involutorial division algebras". *Journal of the Indian Mathematical Society.* 20, (1956), 227–257.
18. "Units of fixed points in involutorial algebras". *Proceedings of International Symposium on Algebraic Number Theory.* Science Council of Japan, Tokyo, 1956, 103–106.
19. "On orthogonal groups". *Nachr. Akad. Wiss. Göttingen Math.-Phys. Kl. II.* (1957), 113–121.
20. "The zeta function and discriminant of division algebra". *Acta Arithmetica.* 5, (1959), 277–288.
21. "Quadratic forms over involutorial division algebras II". *Math. Ann.* 143 (1961), 293–332.
22. "Zeta functions of quadratic forms". *Acta Arithmetica.* 7, (1961), 39–69.
23. "Discontinuous groups". *Nachr. Akad. Wiss. Göttingen Math.-Phys. Kl. II.* 1963, 293–323.
24. "Discontinuous groups II". *Nachr. Akad. Wiss. Göttingen Math.-Phys. Kl. II.* 1964, 145–164.
25. (With S. Raghavan) "On a diophantine inequality concerning quadratic forms". *Nachr. Akad. Wiss. Göttingen Math.-Phys. Kl. II.* 1968, 251–262.
26. "A converse theorem of Siegel". *Prof. Ananda-Rau Memorial Volume.* Publ. Ramanujan Institute, 1, Madras, (1969), 291–296.
27. (With S. Raghavan) "Values of quadratic forms". *Journal of the Indian Mathematical Society.* 34, (1970), 253–257.
28. (With S. Raghavan) "Solvability of a Diophantine inequality in algebraic number fields". *Acta Arithmetica.* 20, (1972), 299–315.
29. "On the analytic theory of quadratic forms". *Acta Arithmetica.* 21, (1972), 423–436.
30. "Theory of numbers". *Journal of Scientific and Industrial Research.* 31, (1972), 459.
31. "Srinivasa Ramanujan, mathematician extraordinary". *Science Today.* Decmeber, (1974).

32. "C. P. Ramanujam: C. P. Ramanujam– A Tribute, Tata Institute of Fundamental Research, *Stud. Math.*. 8, Springer, Berlin (1978), 1–7.
33. (With M. V. Subbarao) "Some generalizations of Ramanujan's sum". *Canadian Journal of Mathematics*. 32, (1980), 1250–1260.
34. "Ramanujan and the congruence properties of partitions". *Proceedings of the Indian Academy of Sciences, Mathematical Sciences*. 89, (1980), 133–157.
35. "The unpublished manuscripts of Srinivasa Ramanujan". *Current Science*. 50, (1981), 203–210.
36. "Remarks on some series considered by Srinivasa Ramanujan". *Journal of the Indian Mathematical Society*. 46, (1982), 107–136.
37. "On Ramanujan's continued fraction". *Acta Arithmetica*. 43, (1984), 209–226.
38. "On the Rogers–Ramanujan continued fraction". *Proceedings of the Indian Academy of Sciences, Mathematical Sciences*. 93, (1984), 67–77.
39. "Ramanujan's continued fraction". *Indian Journal of Pure and Applied Mathematics*. 16, (1985), 695–724.
40. "Srinivasa Ramanujan, 22 December 1887–26 April 1920". *Journal of the Indian Mathematical Society*. 51, (1987), 1–25.
41. "Some applications of Kronecker's limit formula". *Journal of the Indian Mathematical Society*. 52, (1987), 71–89.
42. "Ramanujan's notebooks". *Journal of the Indian Institute of Science*. (1987), Ramanujan Special Issue, 25–32.
43. "Hypergeometric series and continued fractions". *Proceedings of the Indian Academy of Sciences, Mathematical Sciences*. 277–296.
44. "Generalizations of some theorems of Ramanujan". *Journal of Number Theory*. 29, (1988), 118–137.
45. "On some theorems stated by Ramanujan:number theory and related topics, Tata Institute of Fundamental Research". *Stud. Math.* 12, Oxford University Press, 1989, 151–160.
46. "Ramanujan's modular equations'". *Acta Arithmetica*. 53, (1990), 403–420.

Srinivasacharya Raghavan (SR)

1. "Modular forms of degree n and representation by quadratic forms". *Annals of Mathematics* (2). 70, (1959), 446–477.
2. "Modular forms of degree n and representation by quadratic forms". *1960 Contributions to Function Theory* (International Colloquium Function Theory, Bombay, 1960), 181–183, Tata Institute of Fundamental Research, Bombay.
3. "On representation by Hermitian forms". *Acta Arithmetica*. 8, (1962/1963), 33–96.
4. (With Rangachari, S. S. and Sunder Lal) "Algebraic number theory". *Mathematical Pamphlets*, 4. Tata Institute of Fundamental Research, Bombay, (1966).
5. (With Rangachari, S. S.) "On zeta functions of quadratic forms". *Annals of Mathematics*. (2), 85, (1967), 46–57.
6. (With Ramanathan, K. G.) "On a Diophantine inequality concerning quadratic forms". *Nachr. Akad. Wiss. Göttingen Math.-Phys. KI. II.* (1968), 251–262.
7. (With Rangachari, S. S.) "On ternary quadratic forms and modular forms". *Journal of the Indian Mathematical Society* (N. S.). 33, (1969), 187–205.
8. (With Rangachari, S. S.) "On the Siegel formula for ternary skew-Hermitian forms". *Acta Arithmetica*. 16, (1969/1970), 327–345.
9. (With Ramanathan, K. G.) "Values of quadratic forms". *Journal of the Indian Mathematical Society* (N. S.). 34, (1970), no. 3–4, 253–257.
10. (With Ramanathan, K. G.) "Solvability of a Diophantine inequality in algebraic number fields". *Acta Arithmetica*. 20, (1972), 299–315.
11. "On Fourier coefficients of modular forms". *Abh. Math. Sem. Univ.* Hamburg, 38, (1972), 231–237.

12. "On a Diophantine inequality for forms of additive type". Collection of articles dedicated to Carl Ludwig Siegel on the occasion of his seventy-fifth birthday, V, *Acta Arithmetica*. 24, (1973/1974), 499–506.

13. "On an Eisenstein series of degree 3". *Journal of the Indian Mathematical Society* (N. S.). 39, (1975), 103–120.

14. "Bounds for minimal solutions of Diophantine equations". *Nachr. Akad. Wiss. Göttingen Math.-Phys. KI. II.* (1975), no. 9, 109–114.

15. "Cusp forms of degree 2 and weight 3". *Mathematical Annals*. 224, (1976), no. 2, 149–156.

16. "Values of quadratic forms". *Comm. Pure Appl. Math*. 30, (1977), no. 3, 273–281.

17. "Singular modular forms of degree *s*: C. P. Ramanujam – a tribute". Tata Institute of Fundamental Research, *Studies in Mathematics*. 8, Springer, (1978), 263–272.

18. (With Rangachari, S. S.) "Poisson formulae of Hecke type". *Geometry and Analysis*, Indian Academy of Science, Bangalore, 129–149.

19. (With S. G. Dani) "Orbits of Euclidean frames under discrete linear groups". *Israel Journal of Mathematics*. 36, (1980), no. 3–4, 300–320.

20. "Values of indefinite quadratic forms". *Journal of the Indian Mathematical Society* (N. S.). 44, (1980), no. 1–4, 1–21.

21. "Estimates of coefficients of modular forms and generalized modular relations". Automorphic forms, representation theory and arithmetic (Bombay, 1979), 247–254, Tata Institute of Fundamental Research, *Studies in Mathematics*. 10, Springer, (1981).

22. (With Rangachari, S. S.) "Poisson formulae of Hecke type". *Proceedings of the Indian Academy of Sciences, Mathematical Science*. 90, (1981), no. 2, 129–149.

23. "On Ramanujan and Dirichlet series with Euler products". *Glasgow Mathematical Journal*. 25, (1984), no. 2, 203–206.

24. (With Cook, R. J.) "Indefinite quadratic polynomials of small signature". *Montash. Math*. 97, (1984), no. 3, 169–176.

25. "A duality for representation by quadratic forms". *Abh. Math. Sem. Univ*. Hamburg, 54, (1984), 83–90.

26. "On estimates for integral solutions of linear inequalities". *Proceedings of the Indian Academy of Sciences, Mathematical Science*. 93, (1984), no. 2–3, 147–160.

27. "On certain identities due to Ramanujan". *Quarterly Journal of Mathematics*. Oxford, Ser. (2), 37, (1986), no. 146, 221–229.

28. (With Cook, R. J.) "On positive definite quadratic polynomials". *Acta Arithmetica*. 45, (1986), no. 4, 319–328.

29. (With Cook, R. J.) "Positive values of indefinite quadratic forms". *Mathematika*. 33, (1986), no. 1, 164–169.

30. (With Cook, R. J.) "Small independent zeros of quadratic forms". *Mathematical Proceedings of the Cambridge Philosophical Society*. 102 (1987), no. 1, 5–16.

31. "Impact of Ramanujan's work on modern mathematics". Srinivasa Ramanujan Centenary 1987, *Journal Indian Institute of Science*. 1987, Special Issue, 45–53.

32. "On Ramanujan's modular identities". *Proceedings of the Indian Academy of Sciences, Mathematical Science*. 97, (1987), no. 1–3, 263–276, (1988).

33. "Estimation of Fourier coefficients of Siegel modular forms". *Journal of the Indian Mathematical Society* (N. S.). 52, (1987), 23–37, (1988).

34. (With Böcherer, S.) "On Fourier coefficients of Siegel modular forms". *J. Reine Angew. Math*. 384, (1988), 80–101.

35. "Euler products, modular identities and elliptic integrals in Ramanujan's manuscripts I". *Ramanujan Revisited* (Urbana-Champaign, III, 1987), 33, 345, Academic Press, Boston, MA, 1988.

36. (With R. Weissauer) "Estimates for Fourier coefficients of cusp forms". *Number Theory and Dynamical Systems* (York, 1987), 87–102, *London Mathematical Society Lecture Note Series*. 134, Cambridge University Press, Cambridge, (1989).

37. (With Rangachari, S. S.) "On Ramanujan's elliptic integrals and modular identities". *Number Theory and Related Topics* (Bombay 1988), 119–149, Tata Institute of Fundamental Research, *Studies in Mathematics*. 12, TIFR, Bombay, (1989).

38. (With J. Sengupta) "A Dirichlet series for Hermitian modular forms of degree 2". *Acta Arithmetica.* 58, (1991), no. 2, 181–201.
39. "Professor K. G. Ramanathan (1920–1992)". *Acta Arithmetica.* 64, (1993), no. 1, i, 1–6.
40. "A canonical anti-isomorphism of matrix Hecke rings". *Algebra i Analiz.* 5, (1993), no. 2, 211–217; Translation in *St. Petersberg Math. Journal.* 5, (1994), no. 2, 407–413.
41. (With J. Sengupta) "On Fourier coefficients of Maass cusp forms in 3-dimensional hyperbolic space". K. G. Ramanathan Memorial Issue. *Proceedings of the Indian Academy of Sciences, Mathematical Sciences.* 104, (1994), no. 1, 77–92.
42. (With J. Sengupta) "On Fourier coefficients of Maass cusp forms in 3-dimensional hyperbolic space". *Trudy Mat. Inst. Steklov.* 207 (1994), 275–282; Translation in *Proc. Steklov Inst. Math.* 1995, no. 6, (207), 251–257.
43. "The cakravāla method". *Current Science.* 71, (1996), no. 6, 490–493.
44. (With Chan, Wai-kiu and Kim, Myung-Hwan) "Ternary universal integral quadratic forms over real quadratic fields". *Japan Journal of Mathematics* (N. S.). 22, (1996), no. 2, 263–273.
45. (With Kim, B. M. and Kim, M-H) "2-universal positive definite integral quinary diagonal quadratic forms". International Symposium on Number Theory (Madras, 1996). *Ramanujan Journal1.* (1997), no. 4, 333–337.
46. "Glimpses of Ramanujan's work". *Ramanujan Visiting Lectures.* 1–16, Technical Report, 4, Madurai-Kamraj University, Madurai, [1997].

K. Ramachandra (KR)

Publications made in the twentieth century only have been listed.

1. "Some applications of Kronecker's limit formula". *Annals of Mathematics* (2). 80. (1964), 104–148.
2. "On the units of cyclotomic fields". *Acta Arithmetica.* 12, (1966/1967), 165–173.
3. "Approximation of algebraic numbers". *Nachr. Akad. Wiss. Göttingen Math.-Phys. Kl. II.* (1966), 45–52.
4. "Contributions to the theory of transcendental numbers I, II". *Acta Arithmetica.* 14, (1967/1968), 65–72; ibid. 14, (1967/1968), 73–88.
5. "A note on Baker's method". *Journal of the Australian Mathematical Society.* 10, (1969), 197–203.
6. "A note on numbers with a large prime factor". *Journal of the London Mathematical Society* (2). 1, (1969), 303–306.
7. "A lattice-point problem for norm forms in several variables". *Journal of Number Theory.* 1, (1969), 534–555.
8. "On the class of relative Abelian fields". *J. Reine Angew. Math.* 236, (1969), 1–10.
9. "Lectures on transcendental numbers". *The Ramanujan Institute Lecture Notes.* 1, The Ramanujan Institute, Madras, (1969), iii + 73.
10. "A note on numbers with a large prime factor, II". *Journal of the Indian Mathematical Society* (N. S.). 34, (1970), 39–48.
11. "A note on numbers with a large prime factor, III". *Acta Arithmetica.* 19, (1971), 49–62.
12. "A remark on numbers of the form $a^2 - 2b^4$". *Norske Vid. Selsk. Skr.* (Trondheim), (1971), no. 18, 2.
13. "On a discrete mean value theorem for $\zeta(s)$". *Journal of the Indian Mathematical Society* (N. S.). 36, (1972), 307–316.
14. (With T. N. Shorey) "On gaps between numbers with a large prime factor". Collection of articles dedicated to Carl Ludwig Siegel on the occasion of his seventy-fifth birthday, I. *Acta Arithmetica.* 24, (1973), 99–111.
15. "Largest prime factor of the product of k-consecutive integers". Proceedings of the International Conference on Number Theory (Moscow, 1971). *Trudy Mat. Inst. Steklov.* 132, (1973), 77–81.

16. "On the number of Goldbach numbers in small intervals". *Journal of the Indian Mathematical Society* (N. S.). 37, (1973), 157–170.

17. "Application of Baker's theory to two problems considered by Erdös and Selfridge". *Journal of the Indian Mathematical Society* (N. S.). 37, (1973), 25–34.

18. "On the frequency of Titchmarsh's phenomenon for $\zeta(s)$". *Journal of the London Mathematical Society* (2). 8, (1974), 683–690.

19. "A simple proof of the mean fourth power estimate for $\zeta(1/2 + it)$ and $L(1/2 + it, X)$". *Ann. Scuola Norm. Sup. Pisa CI, Sci.* (4), 1, (1974), 81–97.

20. (With Balasubramanian, R.) 'Two remarks on a result of Ramachandra". *Journal of the Indian Mathematical Society* (N. S.). 38, (1974), no. 1–4, 395–397.

21. (With Shorey, T. N. and Tijdeman, R.) "On Grimm's problem relating to factorisation of a block of consecutive integers". *J. Reine Angew. Math.* 273, (1975), 109–124.

22. "On the zeros of a class of generalized Dirichlet series". *J. Reine Angew. Math.* 273, (1975), 31–40.

23. "Application of a theorem of Montgomery and Vaughan to the zeta-function". *Journal of the London Mathematical Society* (2). 10, (1975), no. 4, 482–486.

24. "On a theorem of Siegel". *Nachr. Akad. Wiss. Göttingen Math.-Phys. KI. II.* (1975), no. 5, 43–47.

25. (With Sukthankar, Neela S.) "On Jutila numbers". *Journal of Pure Applied Algebra.* 6, (1975), no. 3, 219–222.

26. "Some new density estimates for the zeros of the Riemann zeta-function". *Ann. Acad. Sci. Fenn. Ser. A I Math.* 1, (1975), no. 1, 177–182.

27. (With Erdös, P. and Babu, G. Jogesh) "An asymptotic formula in additive number theory". *Acta Arithmetica.* 28, (1975/1976), no. 4, 405–412.

28. (With Balasubramanian, R.) "The place of an identity of Ramanujan in prime number theory". *Proceedings of the Indian Academy of Science.* Sect. A, 83, (1976), no. 4, 156–165.

29. (With Huxley, M. N.) "A note on recent papers of Ramachandra and Huxley". *Journal of Number Theory.* 8, (1976), no. 3, 366–368.

30. "Some problems of analytic number theory". *Acta Arithmetica.* 31, (1976), no. 4, 313–324.

31. (With Shorey, T. N. and Tijdeman, R.) "On Grimm's problem relating to factorisation of a block of consecutive integers, II". *J. Reine Angew. Math.* 288, (1976), 192–201.

32. "On the zeros of a class of generalized Dirichlet series II". *J. Reine Angew. Math.* 289, (1977), 174–180.

33. "Two remarks in prime number theory". *Bull. Soc. Math. France,* 105, (1977), no. 4, 433–437,

34. "On the zeros of the Reimann zeta-function and *L*-series". *Acta Arithmetica.* 34, (1977/78), no. 3, 211–218.

35. (With Balasubramanian, R.) "On the zeros of a class of generalized Dirichlet series III". *Journal of the Indian Mathematical Society* (N. S.). 41, (1977), no. 3–4, 301–315.

36. "On the frequency of Titchmarsh's phenomenon for $\zeta(s)$, II". *Acta Math. Akad. Sci. Hungar.* 30, (1977), no. 1–2, 7–13.

37. (With Balasubramanian, R.) "On the frequency of Titchmarsh's phenomenon for $\zeta(s)$ III". *Proceedings of the Indian Academy of Science.* Sect. A, 86, (1977), no. 4, 341–351.

38. (With Erdös, P. and Babu, G. Jogesh) "An asymptotic formula in additive number theory II". *Journal of the Indian Mathematical Society* (N. S.). 41, (1977), no. 3–4, 281–291.

39. (With Balasubramanian, R.) "On the zeros of a class of generalized Dirichlet series IV". *Journal of the Indian Mathematical Society* (N. S.). 42, (1978), no. 1–4, 135–142.

40. "On the zeros of a class of generalized Dirichlet series V". *J. Reine Angew. Math.* 303/304, (1978), 295–313.

41. "Some remarks on the mean value of the Riemann zeta-function and other Dirichlet series I". *Hardy–Ramanujan Journal.* 1, (1978), 15.

42. "Some current problems in multiplicative number theory". *Mathematics Student.* 46, (1978), no. 1, 1–13.

43. "Some remarks on a theorem of Montgomery and Vaughan". *Journal of Number Theory.* 11, (1979), no. 3, S. Chowla Anniversary Issue, 465–471,

44. (With Balasubramanian, R.) "Some problems of analytic number theory II". *Studia Sci. Math. Hungar.* 14, (1979), no. 1–3, 193–202.
45. (With Balasubramanian, R.) "Effective and non-effective results on certain arithmetical functions". *Journal of Number Theory.* 12, (1980), 10–19.
46. "Some remarks on the mean value of the Riemann zeta-function and other Dirichlet series II". *Hardy–Ramanujan Journal.* 3, (1980), 1–24.
47. "One more proof of Siegel's theorem". *Hardy–Ramanujan Journal.* 3, (1980), 25–40.
48. "Some remarks on the mean value of the Riemann zeta–function and other Dirichlet series III". *Ann. Acad, Sci. Fenn. Ser. A I Math.* 5, (1980), no. 1, 145–158.
49. (With M. J. Narlikar) "Contributions to the Erdös–Szemerédi theory of Sieved integers". *Acta Arithmetica.* 38, (1980/81), no. 2, 157–165.
50. "Progress towards a conjecture on the mean value of the Titchmarsh series". *Recent Progress in Analytic Number Theory.* Vol. 1, (Durham, 1979), 303–318, Academic Press, London-New York, (1981).
51. "Progress towards a conjecture on the mean value of the Titchmarsh Series II". *Hardy–Ramanujan Journal.* 4, (1981), 1–12.
52. (With Balasubramanian, R.) "Some problems of analytic number theory III". *Hardy–Ramanujan Journal.* 4, (1981), 13–40.
53. "On series integrals and continued fractions, I". *Hardy–Ramanujan Journal.* 4, (1981), Suppl., 1–11.
54. "Addendum and corrigendum to my paper: one more proof of Siegel's theorem". *Hardy–Ramanujan Journal,* 3, (1980), 25–40;*Hardy–Ramanujan Journal.* 4, (1981), Suppl. 12.
55. (With Balasubramanian, R.) "On the zeros of a class of generalized Dirichlet series VI". *Ark. Mat.* 19, (1981), no. 2, 239–250.
56. "Viggo Brun (13. 10. 1885 to 15. 8. 1978)". *Mathematics Student.* 49, (1981), no. 1, 87–95.
57. (With Alladi, K., Ram Murthy, M. and Sivaramakrishnan, R.) "Problem Session: number theory". (Mysore, 1981), 170–177. *Lecture Notes in Mathematics.* 938, Springer, 1982.
58. "Mean-value of the Riemann zeta-function and other remarks III". *Hardy–Ramanujan Journal.* 6, (1983), 1–21.
59. (With Srinivasan, S.) "A note to a paper: contributions to the theory of transcendental numbers I, II". [*Acta Arithmetica,* 14, (1967/1968), 65–72; ibid. 14, (1967/1968), 73–88]. by Ramachandra on "Transcendental Numbers". *Hardy–Ramanujan Journal.* 6, (1983), 37–44.
60. "Mean-value of the Riemann zeta-function and other remarks [II]". International Conference on Analytical Methods in Number Theory and Analysis (Moscow, 1981). *Trudy Mat. Inst. Steklov.* 163, (1984), 200–204.
61. "Mean-value of the Riemann zeta-function and other remarks I". *Topics in Classical Number Theory.* Vol. I, II (Budapest, 1981), 1317–1347, *Colloq. Math. Soc. János Bolyai.* 34, North-Holland, Amsterdam, (1984).
62. (With Balasubramanian, R.) "Mean-value of the Riemann zeta-function on the critical Line". *Proceedings of the Indian Academy of Sciences, Mathematical Sciences.* 93, (1984), no. 2–3, 101–107.
63. (With Balasubramanian, R.) "Transcendental numbers and a lemma in combinatorics". *Combinatorics and Applications* (Calcutta, 1982), 57–59, Indian Statistical Institute, Calcutta (1984).
64. (With Balasubramanian, R.) "A hybrid version of a theorem of Ingham number theory". (Ootacamund, 1984), 38–46, *Lecture Notes in Mathematics.* 1122, Springer, Berlin, (1985).
65. (With Balasubramanian, R.) "Progress towards a conjecture on the mean value of Titchmarsh series III". *Acta Arithmetica.* 45, (1986), no. 4, 309–318.
66. (With Balasubramanian, R.) "On an analytic continuation of $\zeta(s)$". *Indian Journal of Pure and Applied Mathematics.* 18, (1987), no. 9, 790–793.
67. (With Sankaranarayanan, A.) "A remark on $\zeta(2n)$". *Indian Journal of Pure and Applied Mathematics.* 18, (1987), no. 10, 891–895.
68. "Srinivasa Ramanujan: the inventor of the circle method". (22. 12. 1887 to 26. 4. 1920). *Journal Math. Phys. Sci.* 21, (1987), no. 6, 545–565.

69. "A remark on $\zeta(1 + i)$". *Hardy–Ramanujan Journal.* 10, (1987), 2–8.
70. "Srinivasa Ramanujan: the inventor of the circle method". (22. 12. 1887 to 26. 4. 1920). Inaugural address. *Hardy–Ramanujan Journal.* 10, (1987), 9–24.
71. (With Balasubramanian, R.) "On the number of integers n such that nd $(n) <$ or $= x$". *Acta Arithmetica.* 49, (1988), no. 4, 313–322.
72. (With Balasubramanian, R. and Subbarao, M. V.) "On the error function in the asymptotic formula for the counting function of k-full numbers". *Acta Arithmetica.* 50, (1988), no. 2, 107–110.
73. (With Balasubramanian, R.) "On the frequency of Titchmarsh's phenomenon for $\zeta(s)$, V". *Ark. Mat.* 26, (1988), no. 1, 13–20.
74. (With Balasubramanian, R.) "On square-free numbers". Proceedings of the Ramanujan Centennial International Conference (Annamalainagar, 1987), 27–30, RMS Publ., 1. *Ramanujan Math. Soc.* Annamalainagar, (1988).
75. (With Balasubramanian, R.) "Some local convexity theorems for the zeta-function-like analytic functions". *Hardy–Ramanujan Journal.* 11, (1988), 1–12 (1989).
76. "On the frequency of Titchmarsh's phenomenon for $\zeta(s)$, VII". *Ann. Acad. Sci. Fenn. Ser. A I Math.* 14, (1989), no. 1, 27–40.
77. "An application of Borel–Carathéodory theorem". *Journal of the Ramanujan Mathematical Society.* 4, (1989), no. 1, 45–52.
78. (With Volovich, I. V.) "A generalization of the Riemann zeta-function". *Proceedings of the Indian Academy of Sciences, Mathematical Sciences.* 99, (1989), no. 2, 155–162.
79. (With Sankaranarayanan, A.) "Omega-theorems for the Hurwitz zeta-function". *Arch. Math.* (Basel), 53, (1989), no. 5, 469–481.
80. "Titchmarsh series". *Théorie des nombres* (Quebec, PQ, 1987), 811–814, de Gruyter, Berlin, (1989).
81. (With Balasubramanian, R.) "A lemma in complex function theory, I, II". *Hardy–Ramanujan Journal.* 12, (1989), 1–5, 6–13.
82. "A trivial remark on Goldbach conjecture". *Hardy–Ramanujan Journal.* 12, (1989), 14–19.
83. (With Balasubramanian, R.) "Titchmarsh's phenomenon for $\zeta(s)$". Number Theory and Related Topics (Bombay, 1988), 13–22, Tata Institute of Fundamental Research, *Studies in Mathematics.* 12, TIFR, Bombay, (1989).
84. (With Balasubramanian, R.) "On the frequency of Titchmarsh's phenomenon for $\zeta(s)$, VI". *Acta Arithmetica.* 53, (1990), no. 4, 325–331.
85. (With Balasubramanian, R.) "An alternative approach to a theorem of Tom Meurman". *Acta Arithmetica.* 55, (1990), no. 4, 351–364.
86. (With Balasubramanian, R.) "Proof of some conjectures on the mean-value of Titchmarsh seriesI". *Hardy–Ramanujan Journal.* 13, (1990), 1–20.
87. "Proof of some conjectures on the mean-value of Titchmarsh series with applications to Titchmarsh's phenomenon". *Hardy–Ramanujan Journal.* 13, (1990), 21–27.
88. "On the frequency of Titchmarsh's phenomenon for $\zeta(s)$ IX". *Hardy–Ramanujan Journal.* 13, (1990), 28–33.
89. "A simple proof of Siegel–Tatuzawa theorem". *Bulletin of the Calcutta Mathematical Society.* 82, (1990), no. 3, 222,
90. (With Balasubramanian, R.) "Proof of some conjectures on the mean-value of Titchmarsh series II". *Hardy–Ramanujan Journal.* 14, (1991), 1–20.
91. (With Balasubramanian, R.) "On the zeros of a class of generalized Dirichlet series, VIII, IX". *Hardy–Ramanujan Journal.* 14, 21–33, 34–42.
92. (With Sankaranarayanan, A.) "Note on a paper by H. L. Montgomery, II: extreme values of Riemann zeta-function". [*Comment. Math. Helv.* 52, (1977), no. 4, 511–518]. With an Appendix by the referee. *Acta Arithmetica.* 58, (1991), no. 4, 299–308.
93. (With Sankaranarayanan, A.) "On some theorems of Littlewood and Selberg, II, III". *Ann. Acad. Sci. Fenn. Ser. AI Math.* 16, (1991), no. 1, 131–137, 139–149.
94. "On the zeros of a class of generalized Dirichlet series, VII". *Ann. Acad. Sci. Fenn. Ser. AI Math.* 16, (1991), no. 2, 391–397.

95. (With Sankaranarayanan, A.) "Notes on the Riemann zeta-function". *Journal of the Indian Mathematical Society* (N. S.). 57, (1991), no. 1–4, 67–77.

96. (With Sankaranarayanan, A.) "Note on a paper by H. L. Montgomery" (Omega theorems for the Riemann zeta-function). *"Extreme values of Riemann zeta-function"*. [*Comment. Math. Helv.* 52, (1977), no. 4, 511–518]. *Publ. Inst. Math.* (Beograd) (N. S.), 50(64), (1991), 51–59.

97. (With Sankaranarayanan, A.) "On the frequency of Titchmarsh's phenomenon for $\zeta(s)$, VIII". *Proceedings of the Indian Academy of Sciences, Mathematical Sciences.* 102, (1992), no. 1, 1–12.

98. (With Balasubramanian, R.) "The mean square of the Riemann zeta-function on the line $\sigma = 1$". *Enseign. Math.* (2), 38, (1992), no. 2, 13–25.

99. (With Balasubramanian, R.) "Proof of some conjectures on the mean-value of Titchmarsh series, III". *Proceedings of the Indian Academy of Sciences, Mathematical Sciences.* 102, (1992), no. 2, 83–91.

100. (With Balasubramanian, R.) "On the zeros of a class of generalized Dirichlet series, X". *Indag. Math.* (N. S.). 3, (1992), no. 4, 377–384.

101. "On Riemann zeta-function and allied questions". *Journées Arithmetiqués.* 1991 (Geneva), Astérisque No. 209, (1992), 57–72.

102. (With Balasubramanian, R.) "On the zeros of a class of generalized Dirichlet series, XI". *Proceedings of the Indian Academy of Sciences, Mathematical Sciences.* 102, (1992), no. 3, 225–233.

103. (With Balasubramanian, R.) "On the zeros of $\zeta(s)$ – a [XII]". *Acta Arithmetica.* 63, (1993), no. 2, 183–191.

104. (With Balasubramanian, R.) "On the zeros of $\zeta(s)$ – a [XII]". *Acta Arithmetica.* 63, (1993), no. 4, 359–366.

105. (With Sankaranarayanan, A.) "On some theorems of Littlewood and Selberg, I". *Journal of Number Theory.* 44, (1993), no. 3, 281–291.

106. (With Balasubramanian, R.) "An application of the Hooley–Huxley contour". *Acta Arithmetica.* 65, (1993), no. 1, 45–51.

107. "Application of a theorem of Montgomery and Vaughan to the zeta-function, II". *Journal of the Indian Mathematical Society* (N. S.). 59, (1993), no. 1–4, 1–11.

108. "A brief report on the zeros of a class of generalized Dirichlet series". *Interdisciplinary Studies on Number Theory* (Japanese) (Kyoto, 1992). *Surikaisekikenkyusho Kokyuroku.* No. 837, (1993), 48–56.

109. (With Balasubramanian, R.) "On the zeros of a class of generalized Dirichlet series, XIV". *K. G. Ramanathan Memorial Issue, Proceedings of the Indian Academy of Sciences, Mathematical Sciences.* 104, (1994), no. 1, 167–176.

110. (With Balasubramanian, R.) "On the zeros of a class of generalized Dirichlet series, XV". *Indag. Math.* (N. S.). 5, (1994), no. 2, 129–144.

111. "Some remarks on the mean-value of the Riemann zeta-function and other Dirichlet series, IV". *Journal of the Indian Mathematical Society* (N. S.). 60, (1994), no. 1–4, 107–122.

112. "When is $I_n = \log_{10}(2^n/n)$ close to an Integer?". *Current Science.* 67, (1994), no. 6, 454–456.

113. (With Sankaranarayanan, A.) "On the zeros of a class of generalized Dirichlet series, XVI". *Math. Scand.* 75, (1994), no. 2, 178–184.

114. "A large value theorem for $\zeta(s)$". *Hardy–Ramanujan Journal.* 18, (1995), 1–9.

115. (With Balasubramanian, R.) "On Riemann zeta-function and allied questions II". *Hardy–Ramanujan Journal.* 18, (1995), 10–22.

116. (With Sankaranarayanan, A.) "On some theorems of Littlewood and Selberg, IV". *Acta Arithmetica.* 70, (1995), no. 1, 79–84.

117. "On the mean-value and Omega-theorems for the Riemann zeta-function". *Tata Institute of Fundamental Research Lectures on Mathematics and Physics.* 85, Published for the Tata Institute of Fundamental Research, Bombay; by Springer, 1995. Xiv + 169 pp.

118. "Simplest, quickest and self-contained proof that $\frac{1}{2} < $ or $= \theta < $ or $= 1$ (θ being the least upper bound of the real parts of the zeros of $\zeta(s)$". *Journal of the Indian Mathematical Society* (N. S.). 61, (1995), no. 1–2, 7–12.

119. (With Sankaranarayanan, A.) "A remark on Vinogradov's mean-value theorem". *Journal of Analysis*. 3, (1995), 111–129.

120. (With Sankaranarayanan, A. and Srinivas, K.) "Addendum to K. Ramachandra's paper: some problems of analytic number theory". [*Acta Arithmetica*. 31, (1976), no. 4, 313–324.] *Acta Arithmetica*. 73, (1995), no. 4, 367–371.

121. "On the zeros of $\zeta^{(l)}(s)$ – a (on the zeros of a class of generalized Dirichlet series XVII)". *Proceedings of the Indian Academy of Sciences, Mathematical Sciences*. 105, (1995), no. 3, 273–279.

122. "Little flowers to I. M. Vinogradov". *Trudy Mat. Inst. Steklov*. 207, (1994), 283–285; translation in *Proc. Steklov Inst. Math*. 1995, no. 6 (207), 259–261.

123. (With Sankaranarayanan, A. and Srinivas, K.) "Ramanujan's lattice point problem, prime number theory and other remarks". *Hardy–Ramanujan Journal*. 19, (1996), 2–56.

124. (With Sankaranarayanan, A. and Srinivas, K.) "Problems and results on $\alpha p - \beta q$". *Acta Arithmetica*. 75, (1996), no. 2, 119–131.

125. (With Sankaranarayanan, A.) "Hardy's theorem for zeta-functions of quadratic forms". *Proceedings of the Indian Academy of Sciences, Mathematical Sciences*. 106, (1996), no. 3, 217–226.

126. (With Balasubramanian, R.) "Some local convexity theorems for the zeta-function-like analytic functions, II". *Hardy–Ramanujan Journal*. 20, (1997), 2–11.

127. (With Balasubramanian, R. and Sankaranarayanan, A.) "On the zeros of a class of generalized Dirichlet series, XVIII (a few remarks on Littlewood's theorem and Titchmarsh points)". *Hardy–Ramanujan Journal*. 20, (1997), 12–28.

128. "On the zeros of a class of generalized Dirichlet series, XIX". *Hardy–Ramanujan Journal*. 20, (1997), 29–39.

129. "Professor Paul Erdös (1913–1996), an obituary". *Current Science*. 72, (1997), no. 1, 78–80.

130. "Fractional moments of the Riemann zeta-function". *Acta Arithmetica*. 78, (1997), no. 3, 255–265.

131. (With Sankaranarayanan, A.) "Vinogradov's threeprimes theorem". *Mathematics Student*. 66, (1997), no. 1–4, 27–72.

132. (With Balasubramanian, R.) "Two remarks on a paper: on the distribution of multiplicities of zeros of Riemann zeta-function" [*Czechoslovak Math. J*. 44 (119), (1994), no. 3, 385–404.] by J. Moser: *Bulletin of the Calcutta Mathematical Society*. 89, (1997), no. 3, 199–208.

133. (With Balasubramanian, R.) "Some local convexity theorems for the zeta-function-like analytic functions, III". *Number Theory* (Tiruchirapalli, 1996), 243–256, Contemp. Math., 210, *American Mathematical Society*. Providence, RI, (1998).

134. (With Balasubramanian, R.) "Some remarks on a lemma of A. E. Ingham". Dedicated to Professors Zoltàn Daróczy and Imre Kàtai. Publ. *Math. Debrecen*. 52, (1998), no. 3–4, 281–289.

135. "Many famous conjectures on primes:meagre but precious progress of a deep nature". *Proceedings of the Indian National Science Academy, Part A*. 64, (1998), no. 5, 643–650.

136. "On a method of Davenport and Heilbronn, I". *Hardy–Ramanujan Journal*. 21, (1998), 15.

137. "A remark on Perron's formula". *Journal of the Indian Mathematical Society* (N. S.). 65, (1998), no. 1–4, 145–151.

138. "Many famous conjectures on primes:meagre but precious progress of a deep nature". *Mathematics Student*. 67, (1998), no. 1–4, 187–199.

139. "On the future of Riemann hypothesis". *Current Science*. 77, (1999), no. 7, 951–953.

140. (With Sankaranarayanan, A.) "Notes on the Riemann zeta-function, II". *Acta Arithmetica*. 91, (1999), no. 4, 351–365.

141. (With Balasubramanian, R. and Sankaranarayanan, A.) "Notes on the Riemann zeta-function, III, IV". *Hardy–Ramanujan Journal*. 22, (1999), 23–33, 34–41.

C. P. Ramanujam (CPR)

1. "Cubic forms over algebraic number fields". *Proceedings of the Cambridge Philosophical Society*. 59, (1963), 683–705.
2. "Sums of the *m*-th powers in *p*-adic rings". *Mathematika*. 10, (1963), 137–146.
3. "A note on automorphism groups of algebraic varieties". *Math. Ann*. 156, (1964), 25–33.
4. "On a certain purity theorem". *Journal of the Indian Mathematical Society* (N. S.). 34, (1970), 1–9.
5. "A topological characterization of the affine plane as an algebraic variety". *Ann. of Math*. (2), 94, (1971), 69–88.
6. "Remarks on the Kodaira Vanishing". *Journal of the Indian Mathematical Society* (N. S.). 36, (1972), 41–51.
7. "On geometric interpretation of multiplicity". *Invent. Math*. 22, (1973/74), 63–67.
8. "Supplement to the article 'Remarks on the Kodaira Vanishing'". *Journal of the Indian Mathematical Society*(N. S.). 36, (1972), 41–51." *Journal of the Indian Mathematical Society* (N. S.). 38, (1974), no. 1–4, 1–4.
9. "The invariance of Milnor's number implies the invariance of the topological type". *American Journal of Mathematics*. 98, (1976), no. 1, 67–68.

T. N. Shorey (TNS)

Professor Shorey has many more research publications than listed here. As they were all published in the twenty first century, they have been left out because of the time-constraints of the present project.

1. "On a theorem of Ramachandra". *Acta Arithmetica*. 20, (1972), 215–221.
2. "Algebraic independence of certain numbers in the *p*-adic domain". *Indag. Math*. 34, (1972), 423–435.
3. "p-adic analogue of a theorem of Tijdeman and its applications". *Indag. Math*. 34, 1972, 436–442.
4. (With Ramachandra, K.) "On gaps between numbers with a large prime factor". *Acta Arithmetica*. 24, (1973), 99–111.
5. "On gaps between numbers with a large prime factor II". *Acta Arithmetica*. 25, (1974), 365–373.
6. "Linear forms in the logarithms of algebraic numbers with small coefficients I". *Journal of the Indian Mathematical Society*. 38, (1974), 271–284.
7. "Linear forms in the logarithms of algebraic numbers with small coefficients II". *Journal of the Indian Mathematical Society*. 38, (1974), 285–292.
8. (With Ramachandra, K. and Tijdeman, R.) "On Grimm's problem relating to factorisation of a block of consecutive integers". *Journal Reine Angew. Math*. 273, (1975), 109–124.
9. "Some applications of linear forms in logarithms". *Seminar Delange—Pisot Poitou*. 1975/76, Paris, Exp. 3.
10. "Some applications of linear forms in logarithms". *Seminar Delange—Pisot Poitou*. 1975/76, Paris, Exp. 28.
11. "On linear forms in the logarithms of algebraic numbers". *Acta Arithmetica*. 30, (1976), 27–42.
12. (With Erdös, P.) "On the greatest prime factor $2^p - 1$ and other expressions". *Acta Arithmetica*. 30, (1976), 257–265.
13. (With Tijdeman, R.) "On the greatest prime factors of polynomials at integer points". *Compositio Math*. Scand., 33, (1976), 187–195.
14. (With Tijdeman, R.) "New applications of Diophantine approximations to Diophantine equations". *Math. Scand*. 39, (1976), 5–18.
15. (With Ramachandra, K. and Tijdeman, R.) "On Grimm's problem relating to factorisation of a block of consecutive integers II". *Journal Reine Angew. Math*. 288, (1976), 192–201.

16. (With Van Der Poorten, A. J., Tijdeman, R. and Schinzel, A.) "Applications of the Gel'fond–Baker method to Diophantine equations". *Transcendence Theory: Advances and Applications.* ed. A. Baker and D. W. Masser, Academic Press, London (1977), 59–77.

17. "On the greatest prime factor of $ax^m + by^n$". *Acta Arithmetica.* 36, (1980), 21–25.

18. (With Balasubramanian, R.) "On the equation $a(x^m - 1)/(x - 1) = b(y^n - 1)/(y - 1)$". *Math. Scand.* 46, (1980), no. 2, 177–182.

19. (With Stewart, C. L.) "On divisors of Fermat, Fibonacci, Lucas and Lehmar numbers II". *Journal of the London Mathematical Society.* (2), 23, (1981), 17–23.

20. "The equation $ax^m + by^m = cx^n + dy^n$". *Acta Arithmetica.* 41, (1982), 255–260.

21. (With Parnami, J. C.) "Subsequences of binary recursive sequences". *Acta Arithmetica.* 40, (1982), 193–196.

22. "On the greatest square free factor of members of a binary recursive sequence". *Hardy–Ramanujan Journal.* 6, (1983), 23–26.

23. "Divisors of convergents of a continued fraction". *Journal Number Theory.* 17, (1983), 127–133.

24. (With Stewart, C. L.) "On the equation $ax^{2t} + bx^t y + cy^2 = d$ and pure powers in recurrence sequences". *Math. Scand.* 52, (1983), 24–36.

25. "Applications of linear forms in logarithms to binary recursive sequences". Seminar on Number Theory, Paris 1981/82, *Progr. Math.*, Birkhauser, Boston, (1983), 287–301.

26. "Linear forms in members of a binary recursive sequence". *Acta Arithmetica.* 43, (1984), 317–331.

27. "On the equation $a(x^m - 1)/(x - 1) = b(y^n - 1)/(y - 1)$ II". *Hardy–Ramanujan Journal.* 7, (1984), 1–10.

28. "On the ratio of values of a polynomial". *Proceedings of the Indian Academy of Sciences, Mathematical Sciences.* 93, (1984), 109–116.

29. (With Mignotte, M and Tijdeman, R.) "The distance between terms of an algebraic recurrence sequence". *Journal Reine Angew. Math.* 349 (1984), 63–76.

30. "Perfect powers in values of certain polynomials at integer points". *Mathematical Proceedings of the Cambridge Philosophical Society.* 99, (1986), 195–207.

31. "On the equation $z^q = (x^n - 1)/(x - 1)$". *Indag. Math.* 48, (1986), 345–351.

32. "On the equation $ax^m - by^n = k$". *Indag. Math.* 48, (1986), 353–358.

33. "Integer solutions of some equations". *Current Science.* 55, No. 17, (1986), 815–817.

34. "Perfect powers in products of integers from a block of consecutive integers". *Acta Arithmetica.* 49, (1987), 135–137.

35. "Ramanujan and binary recursive sequences". *Journal of the Indian Mathematical Society.* 52, (1987), 147–157.

36. (With Srinivasan, S.) "Metrical results on square free divisors of convergents of continued fractions". *Bulletin of the London Mathematical Society.* 19, (1987), 135–138.

37. (With Stewart, C. L.) "Pure powers in recurrence sequences and some related Diophantine equations". *Journal of Number Theory.* 27, (1987), 324–352.

38. (With Murty, Ram and Murty, Kumar) "Odd values of Ramanujan τ-function". *Bull. Soc. Math.* France, 115, (1987), 391–395.

39. (With Evertse, J. H., Győry, K. and Tijdeman, R.) "Equal values of binary forms at integral points". *Acta Arithmetica.* 48, (1987), 379–396.

40. (With Győry, K.) "On the denominators of equivalent algebraic numbers". *Indag. Math.* 50, (1988), 29–41.

41. (With Tijdeman, R.) "Perfect powers in arithmetical progression". *Journal of Madras University (Section B).* 51, (1988), 173–180.

42. "Some exponential Diophantine equations". *New Advances in Transcendence Theory.* Ed. A. Baker, Cambridge University Press (1988), 352–365.

43. "Some exponential Diophantine equations II". *Number Theory and Related Topics.* Ed. S. Raghavan, Tata Institute of Fundamental Research, Bombay (1988), 217–229.

44. "Integers with identical digits". *Acta Arithmetica.* 53, (1989), 81–99.

45. (With Tijdeman, R.) "On the number of prime factors of an arithmetical progression". *Journal Sichuan University*. 26, (1989), 72–74.

46. (With Balasubramanian, R. and Waldschmidt, M.) "On the maximal length of two sequences of consecutive integers with the same prime divisors". *Acta Mathematica Hungarica*. 54, (1989), 225–236.

47. (With Tijdeman, R.) "Perfect powers in products of terms in an arithmetical progression". *Compositio Math*. (1990), 307–344.

48. (With Saradha, N.) "On the ratio of two blocks of consecutive integers". *Proceedings of the Indian Academy of Sciences (Mathematical Sciences)*. 100, (1990), 107–132.

49. (With Tijdeman, R.) "On the greatest prime factor of an arithmetical progression (II)". *Acta Arithmetica*. 53, (1990), 499–504.

50. (With Györy, K. and Mignotte, M) "On some arithmetical properties of weighted sums of S-units". *Mathematica Pannonica* 1/2. (1990), 25–43.

51. (With Tijdeman, R.) "On the greatest prime factor of an arithmetical progression". *A Tribute to Paul Erdös*. Ed. A. Baker, B. Bollobas and A. Hajnal, Cambridge University Press, (1990), 25–43.

52. (With Saradha, N.) "The equations $(x + 1) \cdots (x + k) = (y + 1) \cdots (y + mk)$, $m = 3, 4$". *Indag. Math*. N. S. 2, (1991), 489 $\cdots$ 510.

53. (With Saradha, N.) "On the equations $(x + 1) \cdots (x + k) = (y + 1) \cdots (y + mk)$". *Indag. Math*. N. S. 3, (1992), 79–90.

54. (With Saradha, N.) "On the equation $x(x + d) \cdots [x + (k- 1)d] = y(y + d) \cdots [y + (mk - 1)d]$". *Indag. Math*. N. S. 3, (1992), 237–242.

55. (With Tijdeman, R.) "Perfect powers in arithmetical progression (II)". *Compositio Math*. 82, (1992), 107–117.

56. (With Tijdeman, R.) "Perfect powers in products of terms in an arithmetical progression (II)". *Compositio Math*. 82, (1992), 119–136.

57. (With Tijdeman, R.) "Perfect powers in products of terms in an arithmetical progression (III)". *Acta Arithmetica*, 61, (1992), 391–398.

58. (With Tijdeman, R.) "On the number of prime factors of a finite arithmetical progression". *Acta Arithmetica*. 61, (1992), 375–390.

59. (With Tijdeman, R.) "On the greatest prime factor of an arithmetical progression (III)". *Diophantine Approximation and Transcendental Numbers*. Luminy 1990. Ed. Ph. Philippon, Walter de Gruyter, New York, (1992), 275–280.

60. "On the equation $x^l + y^l = 2z^l$ and related problems". *Seminar on Number Theory*. Caen 1992/1993, University of Caen, Exp. VI.

61. (With Balasubramanian, R.) "On the equation $f(x + 1) \cdots f(x + k) = f(y + 1) \cdots f(y + mk)$". *Indag. Math*. N. S. 4, (1993), 257–267.

62. (With Balasubramanian, R.) "Squares in products from a block of consecutive integers". *Acta Arithmetica*. 65, (1994), 213–220.

63. (With Saradha, N.) "On the equation $x(x + d_1) \cdots [x + (k - 1)d_1] = y(y + d_2) \cdots [y + (mk - 1)d_2]$". *Proceedings of the Indian Academy of Sciences (Mathematical Sciences)*. 104, (1994), 1–12.

64. (With Saradha, N. and Tijdeman, R.) "On arithmetic progressions with equal products". *Acta Arithmetica*. 68, (1994), 89–100.

65. "Applications of Baker's theory of linear forms in logarithms to exponential Diophantine Equations'sanalytic number theory, RIMS Kokyuroku, 886, (1994), 48–60, Kyoto University.

66. (With Saradha, N. and Tijdeman, R.) "On arithmetic progressions of equal lengths with equal products". *Mathematical Proceedings of the Cambridge Philosophical Society*. 117, (1995), 193–201.

67. (With Saradha, N. and Tijdeman, R.) "On the equation $x(x + 1) \cdots (x + k - 1) = y(y + d) \cdots [y + (mk- 1) d]m = 1, 2$". *Acta Arithmetica*. 71, (1995), 181–196.

68. "On a conjecture that a product of k consecutive positive integers is never equal to a product of mk consecutive positive integers except for $8.9.10 = 6!$ andrelated problems". *Number Theory*. Paris 1992–1993, Ed. S. David, London Mathematical Society. *Lecture Note Series*. 215, (1995), 231–244.

69. (With Saradha, N. and Tijdeman, R.) "On values of a polynomial at arithmetic progressions with equal products". *Acta Arithmetica*, 72, (1995), 67–76.
70. "Perfect powers in poducts of arithmetical progressions with fixed initial term". *Indag. Math.* N. S. 7, (1996), 521–525.
71. (With Nesterenko, Yu. V.) "Perfect powers in poducts of integers from a block of consecutive integers II". *Acta Arithmetica.* 76, (1996). 191–198.
72. (With Mignotte, M.) "The equations $(x + 1) \cdots (x + k) = (y + 1) \cdots (y + mk)$, $m = 5, 6$". *Indag. Math.* N. S. 7, (1996), 215–225.
73. (With Balasubramanian, R., Langevin, M. and Waldschmidt, M.) "On the maximal length of two sequences of integers in arithmetic progressions with the same prime divisors". *Monatshefte für Mathematik.* 121, (1996), 295–307.
74. "Some applications of Diophantine approximations to Diophantine equations". *Number Theory.* Paris 1993–94, Ed. S. David. London Mathematical Society. *Lecture Note Series.* 235, (1996), 189–198.
75. (With Tijdeman, R.) "Some methods of Erdös applied to finite arithmetic progressions". *The Mathematics of Paul Erdös, Ed. Ronald L. Graham and Jaroslav Nešetřil.* Springer, (1997), 251–267.
76. (With Balasubramanian, R.) "Perfect powers in products of terms in an arithmetical progression (IV)". *Number Theory, Contemporary Mathematics.* 210, (1997), 257–263, American Mathematical Society.
77. (With Hirata-Kohono, Noriko) "On the equation $(x^m - 1)/(x - 1) = y^q$ with x power". *Analytic Number Theory.* Ed. Y. Motohashi, London Mathematical Society. *Lecture Note Series.* 247, (1997), 341–351.
78. (With Tijdeman, R.) "Irrationality criteria for numbers of Mahler's type". *Analytic Number Theory.* Ed. Y. Motohashi, London Mathematical Society. *Lecture Note Series.* 247, 341–351.
79. (With Nesterenko, Yu. V.) "On an equation of goormaghtigh". *Acta Arithmetica.* 83, (1998), 381–389.
80. "Integer solutions of exponential diophantine equations". *Bulletin of Bombay Mathematical Colloquium.* 13, (1998), 1–21.
81. (With Saradha, N.) "The equation $(x^n - 1)/(x - 1) = y^q$ with x square". *Mathematical Proceedings of the Cambridge Philosophical Society.* 125, (1999), 1–19.
82. (With Bugeaud, Mignotte, M. and Roy, Y.) "The equation $(x^n - 1)/(x - 1) = y^q$ has no solution with x square". *Mathematical Proceedings of the Cambridge Philosophical Society.* 127, (1999), 353–372.
83. "The equation $a(x^n - 1)/(x - 1) = by^q$ with $ab > 1$". *Number Theory in Progress.* Volume 1, (1999), Walter de Gruyter, Berlin, 473–485.
84. (With Beukers, F. and Tijdeman, R.) "Irreducibility of polynomials and arithmetic progressions with equal products of terms". *Number Theory in Progress.* Volume 1, (1999), Walter de Gruyter, Berlin, 473–485.
85. "Exponential Diophantine equations involving products of consecutive integers and related equations". *Number Theory.* Ed. R. P. Bambah, V. C. Dumir and R. J. Hans-Gill, Hindustan Book Agency, (1999), 463–s495.

R. Balasubramanian (RB)

Because of the time-constraint of the current project, Publications up to the end of the twentieth century only have been listed here.

1. (With Ramachandra, K.) "Two remarks on a result of Ramachandra". *Journal of the Indian Mathematical Society* (N. S.). 38, (1974), no. 1–4, 395–397.
2. (With Ramachandra, K.) "The place of an identity of Ramanujan in prime number theory". *Proceedings of the Indian Academy of Sciences, Sect. A.* 83, (1976), no. 4, 156–165.
3. (With Ramachandra, K.) "On the zeros of a class of generalized Dirichlet series III". *Journal of the Indian Mathematical Society* (N. S.). 41, (1977), no. 3–4, 301–315.

4. (With Ramachandra, K.) "On the frequency of Titchmarsh's phenomenon for $\zeta(s)$, III". *Proceedings of the Indian Academy of Sciences, Sect. A.* 86, (1977), no. 4, 341–351.

5. "An improvement on a theorem of Titchmarsh on the mean square of $|\zeta(1/2 + it)|$". *Proceedings of the London Mathematical Society.* (3) 36, (1978), no. 3, 540–576.

6. (With Ramachandra, K.) "On the zeros of a class of generalized Dirichlet series IV". *Journal of the Indian Mathematical Society* (N. S.). 42, (1978), no. 1–4, 135–142.

7. "A note on Hurwitz's zeta-function". *Ann. Acad. Sci. Fenn. Ser. A I Math.* 4, (1979), no. 1, 41–44.

8. "Analytic sufficiency conditions for Goldbach's conjecture". *Kyungpook Math. Journal.* 19, (1979), no. 1, 85–105.

9. "On Waring's problem: $g(4) \leq 21$". *Hardy–Ramanujan Journal.* 2, (1979), 31.

10. (With Ramachandra, K.) "Some problems of analytic number theory II". *Studia Sci. Math. Hungar.* 14, (1979), no. 1–3, 193–202.

11. (With Ramachandra, K.) "Effective and non-effective results on certain arithmetical functions". *Journal of Number Theory.* 12, (1980), no. 1, 10–19.

12. (With Shorey, T. N.) "On the equation $a(x^m - 1)/(x - 1) = b(y^n - 1)/(y - 1)$". *Math. Scand.* 46, (1980), no. 2, 177–182.

13. "A note on Dirichlet's L-functions". *Acta Arithmetica.* 38, (1980/81), no. 3, 273–283.

14. (With Ramachandra, K.) "Some problems of analytic number theory III". *Hardy–Ramanujan Journal.* 4, (1981), 13–40.

15. (With Ramachandra, K.) "On the zeros of a class of generalized Dirichlet series VI". *Ark. Mat.* 19, (1981), no. 2, 239–250.

16. (With Murty, M. Ram) "An Ω-theorem for Ramanujan's τ-function". *Invent. Math.* 68, (1982), no. 2, 241–252.

17. (With Ramachandra, K.) "On the zeros of the Riemann zeta-function and L-series II". *Hardy–Ramanujan Journal.* 5, (1982), 1–30.

18. (With Mozzochi, C. J.) "Siegel zeros and the Goldbach problem". *Journal of Number Theory.* 16, (1983), no. 3, 311–332.

19. (With Mozzochi, C. J.) "An improved upper bound for $G(k)$ in Waring's problem for relatively small k". *Acta Arithmetica.* 43, (1984), no. 3, 283–285.

20. (With Ramachandra, K.) "Mean-value of the Riemann zeta-function on the critical line". *Proceedings of the Indian Academy of Sciences, Mathematical Sciences.* 93, (1984), no. 2–3, 101–107.

21. (With Ramachandra, K.) "Transcendental numbers and a lemma in combinatorics". *Combinatorics and Applications* (Calcutta 1982). Indian Statistical Institute, Calcutta, 1984.

22. (With Conrey, J. B. and Heath-Brown, D. R.) "Asymptotic mean square of the product of the Riemann zeta-function and a Dirichlet polynomial". *Journal Reine Angew. Math.* 357, (1985), 161–181.

23. (With Ramachandra, K.) "A hybrid version of a theorem of Ingham". *Number Theory* (Ootacamund, 1984), 38–46, *Lecture Notes in Mathematics.* 1122, Springer, Berlin, (1985).

24. "On Waring's problem: $g(4) \leq 20$". *Hardy–Ramanujan Journal.* 8, (1985), 1–40.

25. (With Ramachandra, K.) "Progress towards a conjecture on the mean value of Titchmarsh series III". *Acta Arithmetica.* 45, (1986), no. 4, 309–318.

26. (With Deshouillers, Jean-Marc and Dress, François) "Problème de Waring pour les bicarrès. I. Schèma de la solution. (French) [Waring's Problem for Biquadrates. I. Sketch of the Solution]". *C. R. Acad. Sci. Paris, Sér. I. Math.* 303, (1986), no. 4, 85–88.

27. (With Deshouillers, Jean-Marc and Dress, François) "Problème de Waring pour les bicarrès. II. Rèsultats Auxiliaires pour le Thèoreme Asymptotique. (French) [Waring's Problem for Biquadrates. II. Auxiliary Results for the Asymptotic Theorem]". *C. R. Acad. Sci. Paris, Sér. I. Math.* 303, (1986), no. 5, 161–163.

28. "Number theory and primality testing". *Workshop on Mathematics of Computer Algorithms* (Madras, 1986), A. 5, 29 pp., IMS Rep., 111, Institute of Mathematical Sciences, Madras, 1986.

29. "On the frequency of Titchmarsh's phenomenon for $\zeta(s)$ IV". *Hardy–Ramanujan Journal.* 9, (1986), 10.

30. (With Ramachandra, K.) "On an analytic continuation of $\zeta(s)$". *Indian Journal of Pure and Applied Mathematics.* 18, (1987), no. 9, 790–796.

31. "A note on a result of Erdös, Sárközy and Sós". *Acta Arithmetica.* 49, (1987), no. 1, 45–53.

32. "The circle method and its implications, Srinivasa Ramanujan Centenary 1987". *Journal Indian Institute of Science.* (1987), Special Issue, 39–44.

33. (With Ramachandra, K.) "On the number of integers n such that $nd(n) \le x$". *Acta Arithmetica.* 49, (1988), no. 4, 313–322.

34. "On the additive completion of squares". *Journal of Number Theory.* 29, (1988), no. 1, 10–12.

35. (With Ramachandra, K. and Subbarao, M. V.) "On the error function in the asymptotic formula for the counting function of k-full numbers". *Acta Arithmetica.* 50, (1988), no. 2, 107–118.

36. (With Ramachandra, K.) "On the frequency of Titchmarsh's phenomenon for $\zeta(s)$ V". *Ark. Mat.* 26, (1988), no. 1, 13–20.

37. (With Adhikari, S. D. and Sankaranarayanan) "On an error term telated to the greatest divisor of n, which is Prime to k". *Indian Journal of Pure and Applied Mathematics.* 19, (1988), no. 9, 830–841.

38. (With Ramachandra, K.) "On square-free numbers". *Proceedings of the Ramanujan Centennial International Conference* (Annamalainagar, 1987), 27–30. RMS Publ. 1. Ramanujan Mathematical Society, Annamalainagar, 1988.

39. (With Ramachandra, K.) "Some local convexity theorems for the zeta-function-like analytic functions". *Hardy–Ramanujan Journal.* 11, (1988), 1–12.

40. (With Shorey, T. N. and Waldschmidt, M.) "On the maximal length of two sequences of consecutive integers with the same prime divisors". *Acta Math. Hungar.* 54, (1989), no. 3–4, 225–236.

41. (With Ramachandra, K.) "A lemma in complex function theory I, II". *Hardy–Ramanujan Journal.* 12, (1989), 1–5, 6–13.

42. (With Adhikari, S. D. and Sankaranarayanan, A.) "An Ω-result related to $r_4(n)$". *Hardy–Ramanujan Journal.* 12, (1989), 20–30.

43. (With Ramachandra, K.) "Titchmarsh's phenomenon for $\zeta(s)$". *Number Theory and Related Topics* (Bombay, 1988). 13–22. Tata Institute of Fundamental Research Studies in Mathematics, 12, TIFR, Bombay, (1989).

44. (With Ramachandra, K.) "On the frequency of Titchmarsh's phenomenon for $\zeta(s)$ VI". *Acta Arithmetica.* 53, (1990), no. 4, 325–331.

45. (With Ramachandra, K.) "An alternative approach to a theorem of Tom Meurman". *Acta Arithmetica.* 55, (1990), no. 4, 351–364.

46. (With Ramachandra, K.) "Proof of some conjectures on the mean-value of Titchmarsh series I". *Hardy–Ramanujan Journal.* 13, (1990), 1–20.

47. (With Murty, M. Ram) "Elliptic pseudoprimes II: Sèminiaire de thèorie des nombres". Paris 1988–1989, 13–25, *Progr. Math.*, 91, Birkhauser, Boston, Boston, MA, 1990.

48. (With Adhikari, S. D.) "A note on a certain error-term". *Arch. Math.* (Basel), 56, (1991), no. 1, 37–40.

49. (With Ramachandra, K.) "Proof of some conjectures on the mean-value of Titchmarsh series II". *Hardy–Ramanujan Journal.* 14, (1991), 1–20.

50. (With Ramachandra, K.) "On the zeros of a class of generalized Dirichlet series, VIII, IX". *Hardy–Ramanujan Journal.* 14, (1991), 21–33, 34–42.

51. (With Soundararajan, K.) "On the additive completion of squares II". *Journal of Number Theory.* 40, (1992), no. 2, 127–129.

52. (With Ramachandra, K. and Sankaranarayanan, A.) "On the frequency of Titchmarsh's phenomenon for $\zeta(s)$, VIII". *Proceedings of the Indian Academy of Sciences, Mathematical Sciences.* 102, (1992), no. 1, 1–12.

53. (With Ivić, A. & Ramachandra, K.) "The mean square of the Riemann zeta-function on the line $\sigma = 1$". *Enseign. Math.* (2), 38, (1992), no. 1–2, 13–25.

54. (With Murty, V. Kumar) "Zeros of Dirichlet L-functions". *Ann. Sci. Ecole Norm. Sup.* (4), 25, (1992), no. 5, 567–615.

55. (With Ramachandra, K.) "On the zeros of a class of generalized Dirichlet series, X". *Indag. Math.* (N. S.), 3, (1992), no. 4, 377–384.

56. (With Ramachandra, K.) "On the zeros of a class of generalized Dirichlet series, XI". *Proceedings of the Indian Academy of Sciences, Mathematical Sciences.* 102, (1992), no. 3, 225–233.

57. (With Ramachandra, K.) "On the zeros of $\zeta(s) - a$, [XII]". *Acta Arithmetica.* 63, (1993), no. 2, 183–191.

58. (With Ramachandra, K.) "On the zeros of $\zeta(s) - a$, [XIII]". *Acta Arithmetica.* 63, (1993), no. 4, 359–366.

59. (With Karunakaran, V. and Ponnusamy, S.) "A proof of Hall's conjecture on starlike mappings". *Journal of the London Mathematical Society.* (2), 48 (1993), no. 2, 278–288.

60. (With Ivić, A. & Ramachandra, K.) "An application of the Hooley–Huxley contour". *Acta Arithmetica.* 65, (1993), no. 1, 45–51.

61. (With Shorey, T. N.) "On the equation $f(x + 1) \ldots f(x + k) = f(y + 1)\ldots f(y + k)$". *Indag. Math.* (N. S.), 4, (1993), no. 3, 257–267.

62. (With Shorey, T. N.) 'Squares in products from a block of consecutive integers". *Acta Arithmetica.* 65, (1993), no. 3, 213–220.

63. (With Raman, Venkatesh and Srinivasaraghavan, G.) "The complexity of finding certain trees in tournaments". *Algorithms and Data Structures* (Montreal, PQ, 1993), 142–150. *Lecture Notes in Computer Science.* 709, Springer, Berlin, 1993.

64. (With Ramachandra, K.) "On the zeros of a class of generalized Dirichlet series, XIV". K. G. Ramanathan Memorial Issue. *Proceedings of the Indian Academy of Sciences, Mathematical Sciences.* 104, (1994), no. 1, 167–176.

65. (With Ramachandra, K.) "On the zeros of a class of generalized Dirichlet series, XV". *Indag. Math.* (N. S.), 5, (1994), no. 2, 129–144.

66. (With Ramachandra, K.) "On Riemann zeta-function and allied questions II". *Hardy–Ramanujan Journal.* 18, (1995), 10–22.

67. (With Soundararajan, K.) "Maximal sets of integers with Distinctdivisors". *Electron. Journal Combin.* Research Paper 22, approx. 5 pp.

68. (With Agarwal, A. K.) "Generalized Gonal numbers and a new class of partitions". *Journal of the Indian Mathematical Society.* (N. S.), 61, (1995), no. 3–4, 153–160.

69. (With Raman, Venkatesh) "Path balance heuristic for self-adjusting binary search trees". Foundations of Software Technology and Theoretical Computer Science (Bangalore, 1995). 338–348. *Lecture Notes in Computer Science.* 1026, Springer, Berlin, 1995.

70. (With Soundararajan, K.) "On a conjecture of R. L. Graham". *Acta Arithmetica.* 75, (1996), no. 1, 1–38.

71. (With Ponnusamy, S.) "An alternate proof of Hall's theorem on a conformal mpping inequality". *Bull. Belg. Math. Soc.* Simon Stevin, 3, (1996), no. 2, 209–213.

72. (With Langevin, M, Shorey, T. N. and Waldschmidt, M.) "On the maximal length of two sequences of integers in arithmetic progressions with the same prime divisors". *Monatsh. Math.* 121, (1996), no. 4, 295–307.

73. (With Nagaraj, S. V.) "Perfect power testing". *Inform. Process. Lett.* 58, (1996), no. 2, 59–63.

74. (With Adhikari, S. D.) "On a question regarding visibility of lattice points". *Mathematika.* 43, (1996), no. 1, 155–158.

75. (With Ramachandra, K.) "Some local convexity theorems for the zeta-function-like analytic functions, II". *Hardy–Ramanujan Journal.* 20, (1997), 2–11.

76. (With Ramachandra, K. and Sankaranarayanan, A.) "On the zeros of a class of generalized Dirichlet series, XVIII (A few remarks on Littlewood's theorem and Titchmarsh points)". *Hardy–Ramanujan Journal.* 20, (1997), 12–28.

77. (With Nagaraj, S. V.) "Density of Carmichael numbers with three prime factors". *Math. Comp.* 66, (1997), no. 220, 1705–1708.

78. (With Raman, Venkatesh and Srinivasaraghavan, G.) "Finding scores in tournaments". *Journal Algorithms.* 24, (1997), no. 2, 380–394.

79. (With Agarwal, A. K.) "*n*-colour partitions with weighted differences equal to minus two". *International Journal of Mathematics and Mathematical Sciences.* 20, (1997), no. 4, 759–768.

80. (With Ramachandra, K.) "Two remarks on a paper: 'On the distribution of multiplicities of zeros of Riemann zeta-function'. *Czechoslovak Math. Journal,* 44, (119), (1994), no. 3, 385–404". *Bulletin of the Calcutta Mathematical Society.* 89, (1997), no. 3, 199–208.

81. (With Ramachandra, K.) "Some local convexity theorems for the zeta-function-like analytic functions, III". *Number Theory.* Tiruchirapalli, 1996, 243–256. *Contemp. Math.* 210. American Mathematical Society, Providence, RI, (1998).

82. (With Shorey, T. N.) "Perfect powers in products of terms in an arithmetical progression IV". *Number Theory.* Tiruchirapalli, 1996. 257–263. *Contemp. Math.* 210. American Mathematical Society, Providence, RI, (1998).

83. (With Ponnusamy, S. and Vuorinen, M.) "Functional inequalities for the quotients of hypergeometric functions". *Journal of Mathematical Analysis and Applications.* 218, (1998), no. 1, 256–268.

84. (With Fellows, Michael R. and Raman, Venkatesh) "An improved fixed-parameter algorithm for vertex cover". *Inform. Process. Lett.* 65, (1998), no. 3, 163–168.

85. (With Koblitz, Neal) "The improbability that an elliptic curve has subexponential discrete log problem under the Menezes–Okamoto–Vanstone algorithm". *Journal Cryptology.* 11, (1998), no. 2, 141–145.

86. (With Ramachandra, K.) "Some remarks on a lemma of A. E. Ingham. Dedicated to Professors Zoltán Daróczy and Imre Kátai". *Publ. Math.* Debrecen. 52, (1998), no. 3–4, 281–289.

87. (With Ponnusamy, S.) "On Ramanujan asymptotic expansions and inequalities for hypergeometric functions". *Proceedings of the Indian Academy of Sciences, Mathematical Sciences.* 108, (1998), no. 2, 95–108.

88. (With Ramana, D. S.) "Atkin's theorem on pseudo-squares". *Publ. Inst. Math.* (Beograd) (N. S.), 63 (77), (1998), 21–25.

89. (With Arasu, K. T. and Evans, A. B.) "A new family of nested row–column designs". *J. Combin. Math. Combin. Comput.* 29, (1999), 139–144.

90. (With Ponnusamy, S.) "Applications of duality principle to integral transforms of analytic functions". *Complex Variables Theory Appl.* 38, (1999), no. 4, 289–305.

91. (With Ramachandra, K., Sankaranarayanan, A. and Srinivas, K.) "Notes on the Riemann zeta-function, III, IV". *Hardy–Ramanujan Journal.* 22, (1999), 23–33, 34–41.

A. Sankaranarayanan (AS)

Because of the time-constraint of the current project, publications up to the end of the twentieth century only have been listed here.

1. "An identity involving Riemann zeta-function". *Indian Journal of Pure and Applied Mathematics.* 18, (1987), 79–800.

2. (With Ramachandra, K.) "A remark on $\zeta(2n)$". *Indian Journal of Pure and Applied Mathematics.* 18, (1987), 891–895.

3. (With Adhikari, S. D. and Balasubramanian, R.) "On an error term related to the greatest divisor of n, which is prime to k". *Indian Journal of Pure and Applied Mathematics.* 19, (1988), 830–841.

4. (With Ramachandra, K.) "Omega-theorems for the Hurwitz zeta-function". *Arch. Math.* 53, (1989), 469–481.

5. (Wth Adhikari, S. D. and Balasubramanian, R.) "On an error term related to $r_4(n)$". *Hardy–Ramanujan Journal.* 12, (1989), 20–30.

6. (With Adhikari, S. D.) "On an error term related to the Jordan totient function $J_k(n)$". *Journal of Number Theory.* 34, (1990), 178–188.

7.   (With Ramachandra, K.) "Notes on the Riemann zeta-function". *Journal of the Indian Mathematical Society*. 57, (1991), 67–77.

8.   (With Ramachandra, K.) "Note on a paper by H. L. Montgomery – I". *Publ. Inst. Math.* 50 (64), (1991), 51–59.

9.   (With Ramachandra, K.) "Note on a paper by H. L. Montgomery – II". *Acta Arithmetica*. 58, (1991), 299–308.

10. (With Ramachandra, K.) "On some theorems of Littlewood and Selberg –II". *Annales Acad. Sci. Fennicae*, Ser. *A1, Mathematica*. 16, (1991), 131–137.

11. (With Ramachandra, K.) "On some theorems of Littlewood and Selberg. III". *Annales Acad. Sci. Fennicae, Ser. A1, Mathematica*. 16, (1991), 139–149.

12. (With Srinivas, K.) "On the papers of Ramachandra and Katai". *Acta Arithmetica*. 62, (1992), 373–382.

13. (With Balasubramanian, R. and Ramachandra, K.) "Titchmarsh's Phenomenon for $\zeta(s)$ – VIII". *Proceedings of the Indian Academy of Sciences, (Mathematical Sciences)*. 102, (1992), 1–12.

14. (With Ramachandra, K.) "On some theorems of Littlewood and Selberg. I". *Journal of Number Theory*. 44, (1993), 281–291.

15. "On the sign changes in the Remainder term of an asymptotic formula for the number of square-free numbers". *Arch. Math.* 60, (1993).

16. (With Srinivas, K.) "Mean-value theorem of the Riemann zeta-function over shorter intervals". *Journal of Number Theory*. 45, (1993), 32–326.

17. (With Ramachandra, K.) "On the zeros of a class of generalised Dirichlet series – XVI". *Math. Scan*. 75, (1994), 178–184.

18. "Zeros of quadratic zeta-functions on the critical line". *Acta Arithmetica*. 69, (1995), 21–38.

19. (With Ramachandra, K.) "A remark on Vinogradov's mean-value theorem". *Journal of Analysis*. 3, (1995), 111–129.

20. (With Ramachandra, K.) "On some theorems of Littlewood and Selberg. IV". *Acta Arithmetica*. 70, (1995), 79–84.

21. (With Ramachandra, K. and Srinivas, K.) "Addendum to Ramachandra's paper, 'some problems of analytic number theory – 1". *Acta Arithmetica*. 73, (1995), 367–371.

22. (With Ramachandra, K. and Srinivas, K.) "Problems and results on $\alpha p - \beta q$". *Acta Arithmetica*. 75, (1995), 119–131.

23. "On a divisor problem related to Epstein zeta-function". *Arch. Math.* 65, (1995), 303–309.

24. (With Ramachandra, K.) "Hardy's theorem for zeta-functions of quadratic forms". *Proceedings of the Indian Academy of Sciences, (Mathematical Sciences)*. 106, (1996), 217–226.

25. (With Ramachandra, K. and Srinivas, K.) "Ramanujan's lattice point problem, prime number theory and other remarks". *Hardy–Ramanujan Journal*. 19, (1996), 2–56.

26. (With Ramachandra, K.) "Vinogradov's three prime theorem". *Mathematics Student*. 66, (1997), 27–72.

27. (With Srinivas, K.) "On a method of Ramachandra and Balasubramanian (on the Abelian group problem)". *Rend. Sem. Mat. Univ. Padova*. 97, (1997), 135–161.

28. (With Balasubramanian, R. and Ramachandra, K.) "On the zeros of a class of generalised Dirichlet series – XVIII". *Hardy–Ramanujan Journal*. 20, (1997), 12–28.

29. "Goldbach problem in polynomial values". *Rend. Circ. Mat. Palermo*. 48, (1999), 243–256.

30. (With Ramachandra, K.) "Notes on the Riemann zeta-function – II". *Acta Arithmetica*. 91, (1999), 351–365.

S. D. Adhikari (SDA)

1.   (With Balasubramanian, R. and Sankaranarayanan, A.) "An Ω-result related to $r_4(n)$". *Hardy–Ramanujan Journal*. 12, (1989), 20–30.

2.   (With Sankaranarayanan, A.) "On an error term related to the Jordan totient function $J_k(n)$". *Journal of Number Theory*. 34, (1990), 178–188.

3.   (With Balasubramanian, R.) "A note on a certain error term". *Arch. Math.* (Basel), 56, (1991), 37–40.

4.   "Omega-results for sums of Fourier coefficients of cusp forms". *Acta Arithmetica*. 57, no. 2, (1991), 83–92.

5. (With Petermann, Y. F. S.) "Lattice points in ellipsoids". *Acta Arithmetica.* 59, no. 4, (1991), 329–338.
6. "Towards the exact nature of a certain error term". *Arch. Math.* (Basel), 58, (1992), 257–264.
7. (With Soundararajan, K.) "Towards the exact nature of a certain error term – II'. *Arch. Math.* (Basel), 59, (1992), 442–449.
8. (With Balasubramanian, R.) "On a question regarding visibility of lattice points". *Mathematika.* 43, (1996), 155–158.
9. "A note on a question of Erdös". *Exposition Math.* 15, no. 4, (1997), 367–371.
10. (With Thangadurai, R.) "A note on sets having the Steinhaus property". *Note di Mathematica.* 16, (1996), no. 1, 77–80. (1998).
11. (With Yong–Gao Chen) "On a question regarding visibility of lattice points". *Acta Arithmetica.* 89, no. 3, (1999), 279–282.

Dipendra Prasad (DP)

1. 'Trilinear Forms for Representations of GL (2) and local ε- factors.': Composito Mathematica, 75, (1990), no. 1, 1–46.
2. (With Gross, Benedict H.) 'Test Vectors for Linear Forms.': Math. Ann., 291, (1991), 343–355.
3. (With Gross, Benedict H.) "On the decomposition of a representation of SO_n when restricted to SO_{n-1}". *Canadian Journal of Mathematics.* 44, (1992), no. 5, 974–1002.
4. "Invariant forms for representations of GL_2 over a local field". *American Journal of Mathematics.* 114, (1992), no. 6, 1317–1363.
5. "On the decomposition of a representation of GL(3) restricted to GL(2) over a p-adic field". *Duke Mathematical Journal.* 69, (1993), no. 1, 167– 177.
6. "Weil representation, howe duality, and the theta correspondence:theta functions from the classical to the modern, 105–127. *CRM Proc. Lecture Notes.* 1, American Mathematical Society, Providence, RI, (1993).
7. "On the local howduality correspondence". *International Mathematical Research Notices.* (1993), no. 11, 279–287.
8. "Bèzout's theorem for Abelian varieties". *Exposition. Math.* 11, (1993), no. 5, 465–467.
9. (With Gross, Benedict H.) "On irreducible representations of $SO_{2n+1} \times SO_{2m}$". *Canadian Journal of Mathematics.* 46, (1994), no. 5, 930–950.
10. "On an extension of a theorem of Tunnell". *Composito Mathematica.* 94, (1994), no. 1, 19–28.
11. "Ribet's theorem: Shimura–Taniyama–Weil implies Fermat". Seminar on Fermat's Last Theorem (Toronto, ON, 1993–1994), 155–177. *CMS Conf. Proc.* 17, American Mathematical Society, Providence, RI, (1995).
12. (With Ramkrishnan, Dinakar) "Lifting orthogonal representations to spin groups and local root numbers". *Proceedings of the Indian Academy of Sciences (Mathematical Sciences).* 105, (1995), no. 3, 259–267.
13. "Some applications of seesaw duality to branching laws". *Math. Ann.*, 304, (1996), no. 1, 1–20.
14. (With Khare, Chandrasekhar) ""Extending local representations to global representations". *J. Math. Kyoto Univ.* 36, (1996), no. 3, 471–480.
15. "A brief survey on the theta correspondence". Number Theory (Tiruchirapalli, 1996), 171–193. *Contemp. Math.* 210, American Mathematical Society, Providence, RI, 1998.
16. "On the self-dual representations of finite groups of Lie type". *Journal Algebra.* 210, (1998), no. 1, 298–310.
17. "Some remarks on representations of a division algebra and of the Galois group of a local field". *Journal of Number Theory.* 74, (1999), no. 1, 73–97.
18. "On the self-dual representation of a p-adic group". *International Mathematical Research Notices.* (1999), no. 8, 443–452.
19. (With Ramkrishnan, Dinakar) "On the global root numbers of $GL(n) \times GL(m)$". Automorphic Forms, Automorphic Representations and Arithmetic (Fort Worth, TX, 1996), 311–330. *Proceedings of the Symposium on Pure Mathematics.* 66, Part 2, American Mathematical Society, Providence, RI, 1999.
20. "Distinguished representations for quadratic extensions". *Composito Math.* 119, (1999), no. 3, 335–345.

Section 4

B. N. Seal (BNS)

1. "The equation of digits; being an elementary application of a principle of numerical grouping to the solution of numerical equation". *Bulletin of the Calcutta Mathematical Society.* 10, (1919), 99–123.

H. Datta (HD)

1. "On some properties of natural numbers". *Bulletin of the Calcutta Mathematical Society.* 10, (1919), 229–238.

S. C. Mitra (SCM)

1. "On the proof of a result given by Ramanujan about the complex multiplication of elliptic function". *Bulletin of the Calcutta Mathematical Society.* 24, (1932), 135–136.

D. P. Banerjee (DPB)

1. "On solution of the 'easier' waring problem". *Bulletin of the Calcutta Mathematical Society.* 34, (1942), 197–199.
2. "Congruence poperties of Ramanujan's function $\tau(n)$". *Journal of the London Mathematical Society.* 17, (1942), 144–145.
3. "On the new congruence properties of the arithmetic function $\tau(n)$". *Proceedings of the National Academy of Sciences, India, Section A.* 12, (1942), 149–150.
4. "On the rational solution of the Diophantine equation $ax^n - by^n = k$". *Proceedings of the Benares Mathematical Society* (N. S.). 5, (1943), 29–30.
5. "On some formulae in analytic theory of numbers". *Bulletin of the Calcutta Mathematical Society.* 36, (1944), 49–50.
6. "On sme formulae in analytic theory of numbers II". *Bulletin of the Calcutta Mathematical Society.* 36, (1944), 107–108.
7. "On the application of the congruence property of Ramanujan's function to certain quaternary form". *Bulletin of the Calcutta Mathematical Society.* 37, (1945), 24–26.
8. "On a theorem in the theory of partition". *Bulletin of the Calcutta Mathematical Society.* 37, (1945), 113–114.
9. "On the divisors of numbers". *Bulletin of the Calcutta Mathematical Society.* 39, (1947), 57–58.
10. "On the self-inverse module". *The Mathematics Student.* 15, (1947), 17–18.
11. "On some identities in the theory of partitions". *Proceedings of the National Academy of Sciences, India, Section A.* 34, (1964), 68–73.

D. B. Lahiri (DBL)

1. "On Ramanujan's function $\tau(n)$ and the divisor function $\sigma_k(n)$ – I". *Bulletin of the Calcutta Mathematical Society.* 38, (1946), 193–206.
2. "On a type of series involving the partition function with applications to certain congruence relations". *Bulletin of the Calcutta Mathematical Society.* 38, (1946), 125–132.
3. "On Ramanujan's function $\tau(n)$ and the divisor function $\sigma_k(n)$ – II". *Bulletin of the Calcutta Mathematical Society.* 39, (1947), 33–52.
4. (With Bambah, R. P., Chowla, S. and Gupta, H.) "Congruence properties of Ramanujan's function $T(n)$". *Quarterly Journal of Mathematics.* Oxford, Ser. 18, (1947), 143–146.
5. "Some non-Ramanujan congruence properties of the partition function". *Proceedings of the National Institute of Sciences, India.* (1948), 337–338.

6. "Further non-Ramanujan congruence properties of the partition function". *Science and Culture.* 14, (1949), 336–337.

7. "Congruence for the Fourier coefficients of the modular invariant tau". *Proceedings of the National Institute of Sciences, India, Part A.* 32, (1966), 95–103.

8. "Identities connecting the partition divisor and Ramanujan's functions". *Proceedings of the National Institute of Sciences, India, Part A.* 34 Suppl. 1, (1968), 96–103.

9. "Identities connecting elementary divisor functions of different degrees and allied congruences". *Math. Scand.* 24, (1969), 102–110.

10. "Some arithmetical identities for Ramanujan's and divisor functions". *Bulletin of the Australian Mathematical Society, I.* (1969), 307–314.

11. "Some restricted partition functions: congruences modulo (7)". *Transactions of the American Mathematical Society.* 140, (1969), 475–484.

12. "Some restricted partition functions: congruences modulo (5)". *Journal of the Australian Mathematical Society.* 9, (1969), 424–432.

13. "Some congruences for the elementary divisor functions". *American Mathematical Monthly.* 76, (1969), 395–397.

14. "Some restricted partition functions: congruences modulo (3)". *Pacific Journal of Mathematics.* 28, (1969), 575–581.

15. "Some restricted partition functions: congruences modulo (2)". *Transactions of the American Mathematical Society.* 147, (1970), 271–278.

16. "Some restricted partition functions: congruences modulo (3)". *Journal of the Australian Mathematical Society.* 11, (1970), 82–90.

17. "Some restricted partition functions: congruences modulo (11)". *Pacific Journal of Mathematics.* 38, (1971), 103–116.

18. "Hypo-multiplicative number-theoretic functions". *Aequationes Math.* 9, (1973), 184–192.

L. G. Sathe (LGS)

1. "On a problem of Hardy (I)". *Journal of the Indian Mathematical Society* (N. S.). 17, (1954), 63–82.

2. "On a problem of Hardy (II)". *Journal of the Indian Mathematical Society* (N. S.). 17, (1954), 83–141.

3. "On a problem of Hardy (III)". *Journal of the Indian Mathematical Society* (N. S.). 18, (1954), 27–42.

4. "On a problem of Hardy (IV)". *Journal of the Indian Mathematical Society* (N. S.). 18, (1954), 43–81.

T. N. Sinha (TNS)

1. "Some systems of Diophantine equations of the Terry–Escott type". *Journal of the Indian Mathematical Society* (N. S.). 30 (1), (1966), 15–26.

2. "Integer solutions of the equations $a_i x_i = 0$". *The Mathematics Student.* 39, (1971), 376–378.

A. N. Sinha (ANS)

1. "On squaring the numbers". *The Mathematics Education.* 1 (4), (1967), 129–132.

M. R. Iyer (MRI)

1. "Identities involving generalized Fibonacci numbers". *Fibonacci Quarterly Journal.* 7 (1), (1969), 66–72.

2. "Sums involving fibonacci numbers". *Fibonacci Quarterly Journal.* 7 (1), (1969), 92–98.

3. "Some results on Fibonacci quaternions". *Fibonacci Quarterly Journal.* 7 (1), (1969), 201–210.

S. A. N. Moorthy (SANM)

1. "A theorem on properties of cubes of natural numbers". *The Mathematics Education.* 6 (3), (1972), 87.
2. "Some results associated with the unique representation of a positive integer as sum of different pwers of two". *The Mathematics Education.* 9 (1), (1975), 9–12.
3. "On Primes and their Distribution". *The Mathematics Education.* 9 (2), (1975), 41.

D. N. Singh (DNS)

1. "A quadratic Diophantine equation". *Journal of the Bihar Mathematical Society.* 12, (1989), 115–116.
2. "Congruence considerations in Diophantine equations". *Journal of the Bihar Mathematical Society.* 13, (1990), 64–66.

J. Choubey (JC)

1. "On convolution in context of number theory". *The Mathematics Education.* 25 (3), (1991), 158–161.
2. "On ideal and non-ideal solutions of the Terry–Escott problem". *The Mathematics Education.* 25 (4), (1991), 246–250.
3. "Method of constructing special solutions of Diophantine equations". *The Mathematics Education.* 28 (1), (1994), 22–25.

References

1. Alladi, K. *Ramanujan's Place in the World of Mathematics: Essays Providing a Comparative Study.* Springer (2012).
2. Andrews, G. E. *Ramanujan Revisited Proceedings of the Centenary Conference.* Academic Press (1988).
3. Burton, David M. *Elementary Number Theory.* McGraw Hill (1976).
4. Grosswald, Emil. *Representations of Integers as Sum of Squares.* 3rd ed. Springer, New York, (1985).
5. Jones, Gareth A. and Jones, J. Mary. *Elementary Number Theory.* Springer Undergraduate Mathematics Series (2005).
6. Knopp, Marvin, I. *Modular Functions in Analytic Number Theory.* 2nd ed., American Mathematical Society, Chelsea Publications, (1993).
7. Murty, K. Ram and Murty, V. Kumar. *The Mathematical Legacy of Srinivasa Ramanujan.* Springer (2012).
8. Narkiewicz, W. *Rational Number Theory in the 20th Century.* Springer (2012).
9. Rao, K. Srinivasa. *Srinivasa Ramanujan: A Mathematical Genius.* East West Books, Madras (2005).
10. Serre, Jean-Pierre. *A Course in Arithmetic.* Springer, New York, (1973).

Printed in the United States
by Baker & Taylor Publisher Services

Printed in the United States
by Baker & Taylor Publisher Services